CHEMISTRY RESEARCH AND APPLICATIONS

THE DEVELOPMENT OF ORGANIC PHOSPHORUS CHEMISTRY

DISCOVERIES OF THE NINETEENTH CENTURY

CHEMISTRY RESEARCH AND APPLICATIONS

Additional books in this series can be found on Nova's website under the Series tab.

Additional e-books in this series can be found on Nova's website under the e-book tab.

CHEMISTRY RESEARCH AND APPLICATIONS

THE DEVELOPMENT OF ORGANIC PHOSPHORUS CHEMISTRY

DISCOVERIES OF THE NINETEENTH CENTURY

LOUIS DUBOSE QUIN

Professor Emeritus

Duke University and the University of Massachusetts

NOVINKA

New York

For permission to use material from this book please contact us:
Telephone 631-231-7269; Fax 631-231-8175
Web Site: http://www.novapublishers.com

Additional color graphics may be available in the e-book version of this book.

LIBRARY OF CONGRESS CATALOGING-IN-PUBLICATION DATA

The development of organic phosphorus chemistry : discoveries of the nineteenth century / editor, Louis DuBose Quin (Duke Univ. and Univ.of Massachusetts, US).
pages cm. -- (Chemistry research and applications)
Includes index.
ISBN: 978-1-63321-333-3 (softcover)
1. Organometallic chemistry--History--19th century. 2. Phosphorus compounds. 3. Organophosphorus compounds. 4. Chemistry, Organic--History--19th century. 5. Chemists. I. Quin, Louis D., 1928- editor.
QD412.P1D48 2014
547'.07--dc23

2014022549

Published by Nova Science Publishers, Inc. † New York

Dedicated to the Memory of
Dr. Gyöngyi Szakal Quin
October 14, 1939 – August 8, 2011

The next major figure to take over phosphorus chemistry was August ichaelis; his research on carbon-phosphorus compounds spanned the period 873-1916 and resulted in many major discoveries. While most of his work as performed with aromatic compounds, he is best known today as the riginator of two still-useful methods for the synthesis of alkyl phosphonic cid derivatives. Chapter 3 is devoted entirely to a presentation of his xtensive record of accomplishments.

Other chemists became attracted to this new field in the last quarter of the nineteenth century and were successful in opening up new pathways and in expanding knowledge around the discoveries of Hofmann and Michaelis. Their work is described in Chapter 4.

The foundations of the field of organophosphorus chemistry were therefore in place when the twentieth century opened. The expansion of the field thereafter was dramatic, and became widely international in scope. The magnitude of this research effort can easily be appreciated by considering the highly important monograph by Gennady M. Kosolapoff, *Organophosphorus Compounds* (John Wiley & Sons), published in 1950 as the first book on the subject in English. This monograph, with its extensive compilation of references to the early literature, was of major help in the preparation of the present book. The field has grown to an enormous size, with many hundreds of thousands of known compounds. Now, in 2014, there are many books on the subject, some general in scope, others on specific aspects.

It was curiosity that led to the first synthesis of an organophosphorus compound, and it was curiosity on the part of the present author that led to the writing of this book. I have worked in the field for over half a century, and developed high appreciation for the work of the early chemists. They accomplished much without the facilities, instruments, and techniques of modern research, and my curiosity led to a study of how it started and developed. This book is a tribute to the brilliance, skill, and even bravery of the nineteenth century chemists.

Finally, I thank Mary Risher Tatum, A.S.H., for her interest and support during the preparation of this book.

Louis DuBose Quin
Professor Emeritus,
Duke University and University of Massachusetts
Durham, NC 27705, May 1, 2014
Tel: 919 384 2412
E-mail: ldquin27@gmail.com

CONTENTS

PREFACE

It was the curiosity of a nineteenth century chemist that led discovery of the first organic compound to have carbon bonded direc phosphorus. The natural world had been the major source of or compounds in the early days of organic chemistry, but such phosph compounds were not known until 1959 to occur in Nature. This new synth field could not have gotten off to a worse start. In 1845, the French cher Paul Thenard carried out a reaction of calcium phosphide with methyl chlor in sealed glass tubes at a high temperature. Some tubes exploded, and tho that survived contained horribly smelly compounds, frequently catching c fire when being opened and exposed to the atmosphere. Nevertheless, Thenar was able to separate the liquid compound trimethylphosphine ($(CH_3)_3P$) from the reaction products and prove its composition by elemental analysis. Thus trimethylphosphine stands as the first organic compound to contain a C-P bond; today it is a very important ligand in coordination compounds. It was, however, not a very good representative of the new field; it had a strong smell and readily reacted with oxygen on exposure to the atmosphere. This work is described in more detail in Chapter 1. It was ten years before anyone made any effort to exploit this lead into a new world of chemistry, and even then, in 1855, the next experiments either repeated Thenard's reaction or made modifications of it that produced a second phosphine, triethylphosphine. However, one of the researchers to conduct such experiments was the brilliant August Wilhelm Hoffman, who became one of the most accomplished of the many outstanding nineteenth century chemists. He took up the challenge of expanding this new field and proceeded to show that phosphines could be synthesized by more attractive conditions, and that new compounds (phosphine oxides and phosphonic acids, among several other types) could be made that were crystalline, odorless derivatives. Most of his results were published in the period 1861-1873, when he became recognized as the leader of this new field. His numerous accomplishments are the subject of Chapter 2.

Chapter 1

The Birth of the Field of Organic Phosphorus Chemistry

Abstract

The first compound with a carbon-phosphorus bond was synthesized by Paul Thenard in 1845. The synthesis was confirmed in 1857, and after that the field began to grow, with pioneering work by August Wilhelm Hofmann and August Michaelis. Much was accomplished in spite of restrictions on experimental techniques and apparatus. The early work was performed before and during the development of the structural theory of organic chemistry, which was not in wide use until about 1870-1875.

1.1. Introduction

Organophosphorus chemistry today is of course a very large and complex field, with many commercial applications of compounds with various functional groups. It is seldom, if ever, asked nowadays how it all began. Indeed, it was a difficult beginning, the very first compound to contain a carbon-phosphorus bond being the hard to handle trimethylphosphine, with a terribly unpleasant smell. That occurred in 1845; no other types of compounds were reported until August Wilhelm Hofmann began his famous series of experiments in 1855, which lasted until 1873 and resulted in the creation of many functional groups that remain the basis of much of phosphorus chemistry today. The second major series of investigations was started by August Michaelis in 1873, again resulting in highly significant discoveries over a period of 42 years. It was during the 1870's that other chemists began to make contributions to the field, some of significance, but most of these authors published only one or two papers; until the end of the century, none of these contributors achieved a high level of prominence in this field.

The treatment of the subject in this book ends with the nineteenth century, by which time it was on very firm ground and began to attract greater interest and research activity on a wider international basis. The next major figure to emerge in the early twentieth century, was Aleksandr E. Arbuzov in Russia, followed by many others of prominence. But that is another story and well covered in many monographs on phosphorus chemistry since 1950, when Gennady Kosolapoff published the first book in English [1].

The structure of the present book is developed along the lines mentioned in this introduction. The earliest, very limited discoveries are presented in detail in this Chapter. Chapter Two is devoted entirely to a summary of the accomplishments of August Hofmann, which are so great that he may be considered to be the Founding Father of the field [2]. In Chapter Three, the massive and enormously important work of August Michaelis is presented. The contributions of other chemists in the period 1870-1900 are described in Chapter Four.

The coverage in this book is entirely based on compounds containing the carbon-phosphorus bond. This is not to say that compounds containing phosphorus groups linked through oxygen to carbon are of no importance. Indeed they are very important and their vital role in life processes and in commerce are well known. The structures being referred to are esters, amides, thio, or chloro derivatives of phosphoric acid, phosphorous acid, H-phosphonic acids ($H\text{-}P(O)(OH)_2$), and hypophosphorous acid ($H_2P(O)(OH)$). The first work with such compounds actually preceded that on the C-P compounds; in 1820 J.-P. Lassaigne in France [3] reacted ethyl alcohol with polyphosphoric acid (present in syrupy phosphoric acid, which he used) presumably giving phosphate esters. However, no data on isolated compounds were given, and the paper is of historical importance only. Three papers in 1833 also reported on the reaction of syrupy phosphoric acid with ethyl alcohol. In this work [4], monoethyl phosphate was isolated as the barium salt, and identified by elemental analysis. Numerous papers were published on the derivatives of the phosphorus acids during the nineteenth century, again forming a good basis for work that continued into the twentieth century. This work does not lend itself to a presentation such as given in this book for carbon-phosphorus compounds, and is not included. For those interested in this aspect of phosphorus chemistry, detailed discussions, covering all the work up to 1950, may be found in the Kosolapoff book [1].

1.2. Experimental Work with Phosphorus Compounds in the Nineteenth Century

To appreciate the greatness of the early experimental work in organic chemistry, one needs to consider the limitations in apparatus and techniques available in the research laboratory. The early techniques for conducting chemical reactions are for the most part obsolete, or at least unappealing. Some of the early reactions attempted were conducted in sealed glass tubes at high temperatures, some at 250-300^0. Explosions and fires on opening the tubes were reported frequently with some types of phosphorus compounds. Solvents for reaction media were not generally used; it was more common just to mix reactants, solids or liquids, and heat them. The first report of a technique for conducting a reaction in a solvent with a water-cooled condenser set in the reflux position was not made until 1836 [5]; mechanical stirring in flasks was well off into the future.

Separations of mixtures was accomplished by extraction, fractional distillation or crystallization, which are still of value in the organic chemical laboratory. No chromatographic techniques were known in the nineteenth century. After purification, the compounds were subjected to the only method available, elemental analysis, that gave information on molecular composition. However, analysis for phosphorus was frequently unreliable and very few papers report phosphorus percentage data. Methods for measuring molecular weight by boiling or freezing phenomena were available but seldom employed. Compounds were characterized as at present by their melting or boiling points, and occasionally by their density or refractive index. Chemical properties were especially useful guides to developing structural formulas. Titrations of acids or bases were in use to obtain equivalent weights. Descriptions of syntheses in the early literature were usually lacking in details of procedure and apparatus. Yields were seldom given, making it difficult to evaluate the efficiency of a process.

Table 1.1. Names of Common Phosphorus Functional Groups

A. Three-Coordinate Functions	
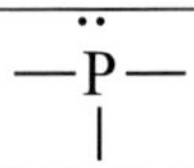	
RPH_2	Alkylphosphine (also phosphane, but not used in this book)
R_2PH	Dialkylphosphine
R_3P	Trialkylphosphine
$RPCl_2$	Alkylphosphonous dichloride
R_2PCl	Dialkylphosphinous chloride
$RP(OR)_2$	Dialkyl alkylphosphonite
B. Four-coordinate Functions	

$RP(O)(OH)_2$	Alkylphosphonic acid
$RP(O)(OR)_2$	Dialkyl alkylphosphonate
RPH(O)(OH)	Alkylphosphonous acid
$RP(O)Cl_2$	Alkylphosphonic dichloride
$R_2P(O)(OH)$	Dialkylphosphinic acid
$R_2P(O)(OR)$	Alkyl dialkylphosphinate
$R_2P(O)Cl$	Dialkylphosphinic chloride
$R_2PH(O)$	Dialkylphosphine oxide
$R_3P(O)$	Trialkylphosphine oxide
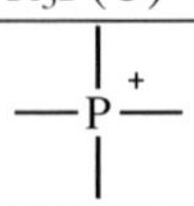	
R_4P+	Phosphonium ion

1.3. The Naming of Organophosphorus Compounds

The nineteenth century chemists discovered many different functional groups containing phosphorus, and additional ones were discovered in later research. Assigning descriptive names to these compounds became a difficulty very early in their study, a problem that was compounded by the fact that the first research was published in German and French journals, and later in

English journals, sometimes with no agreement in the names of the functional groups. It was not until 1979 that the International Union of Pure and Applied Chemistry (IUPAC) put forth a systematic set of rules of nomenclature that is now in universal use [6]. One of the major difficulties had been with the derivatives of phosphorus acids; there are many types, since phosphoryl groups (P=O) can be bonded to hydroxyl and alkoxy groups, halogens, various nitrogen and sulfur functions, etc., and many combinations of these. The problem is especially severe with derivatives of phosphoric acid which must have three of these groups, identical or mixed. However, this is not a matter of concern in the present book, since the emphasis is on phosphorus functions bearing one, two, or three carbon substituents. The concern for readers of this book is to recognize easily the structure-name relations of the basic functional groups that became known to the early workers. In the text, use is made only of the modern names for these groups. The functional groups and associated names to be encountered in the chapters to follow are summarized in Table 1.1. There are two major groups, where phosphorus may have three bonded atoms (said to be three-coordinate or tri-covalent) or four atoms, thus four-coordinate. This system avoids the issue of valence and how to treat the phosphoryl group with its unique type of double bonding. Oxygen is simply taken as one of the four atoms attached to phosphorus. In Table 1.1 and elsewhere, it is written with a conventional double bond symbol, but as is discussed in the text this is not the type of *pi*-bond found in the carbonyl group, etc. Note also in this table that aryl may replace alkyl in all functions, and that many functions have sulfur replacing oxygen.

1.4. THE STATUS OF ORGANIC CHEMICAL THEORY IN THE MID-NINETEENTH CENTURY

We are so accustomed to writing structures for organic compounds showing the atoms connected with lines to indicate electron pair bonds that we do not think about the early days when many organic chemists were conducting important synthetic work with no clear understanding of how molecules were composed. Theories abounded in the first half of the nineteenth century, and are described in excellent fashion in books on historical aspects of organic chemistry published in recent years, e. g. those by A. J. Rocke [7,8] and others. By mid-century, it was understood that groups of atoms such as methyl, ethyl, etc. appeared in compounds and might be considered as a unit that replaced other atoms, especially hydrogen, on some central atom. August Hofmann used this concept in 1850 to explain the existence of primary, secondary, and tertiary amines [9]. He then applied it in 1857 to describe triethylphosphine [10] which he wrote as E_3P, and also in form **1**. Even as late 1872 he was still using such notation, as seen in **2** to represent benzylphosphine.

$$\left.\begin{matrix}E\\E\\E\end{matrix}\right\}P \qquad\qquad \left.\begin{matrix}C_7H_7\\H\\H\end{matrix}\right\}P$$

1 2

These representations are close to showing a structure as we might express it today, but does not specify a connection to P. It was roughly in the period 1855-1860 that the concepts of valence, carbon to carbon bonding, and molecular structure were being developed, by such famous chemists as Kekulé, Erlenmeyer, Butlerov, Couper, Crum Brown and others. This resulted in the Structural Theory for organic compounds, a major triumph in the development of Organic Chemistry. But the theory of the bonding of the atoms, and of the mechanism of reactions, was way off in the future. Alexander Crum Brown in Edinburgh [11] began the representation of structure with lines connecting the atoms in 1864, making use of tetravalence for carbon. His drawings are much like those we write today. Hofmann used the Crum Brown structures to create three-dimensional wooden models that he used in lecture demonstrations [12]. He always expressed his phosphorus compounds in ways easily understandable, but he never used the Crum Brown structures in his publications on his phosphorus research. In fact, it was some years after the Crum Brown structures were published before structural formulas began to appear in the literature. The other great nineteenth century phosphorus chemist, August Michaelis, rarely used them in his long career in phosphorus chemistry (1873-1915). One exception to this statement is the structure **3** he drew to represent a chloro derivative of phenylphosphonous acid [13]. This is one of a very few structures he drew to indicate double bonding in the phosphoryl group.

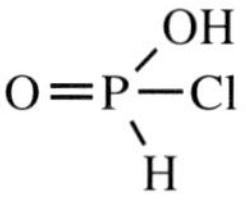

3

Much of his work was based on benzene chemistry, a field placed on solid structural grounds by the famous work of Kekulé who developed the formal cyclohexatriene structure in 1864. It has been stated that the Kekulé structure was accepted by most chemists by 1874 [14]. Michaelis's papers seldom contained structures showing the benzene ring. He simply represented this with the formula C_6H_5 as a substituent on P. Still, he expressed his compounds in clear ways, easily interpreted by present-day chemists.

1.5. THE OBSERVATION OF THE FIRST COMPOUND WITH A CARBON-PHOSPHORUS BOND

To the French chemist Paul Thenard goes the credit for the first observations that a compound composed of carbon, hydrogen and phosphorus could exist. In 1845 [15] he reported on the outcome of a reaction of methyl chloride with calcium phosphide. The reactions were conducted at 180-300°C. in sealed glass tubes, a procedure typical of the times, and without a solvent. There must have been great surprise on opening the first tube and encountering the horrible smell of the products; spontaneous inflammation also occurred. Thenard reported that five products could be observed, three liquid and two solid, but he gives no details on how they were separated or on the quantity of each. He speculated that three of these appeared to be a series of compounds where phosphorus is associated with different amounts of carbon and hydrogen, but insufficient analytical data were given to support this. The compound CH_3PH_2 is a gas at room temperature, and $(CH_3)_2PH$ is a gas above 25°C., and no evidence suggests these were observed, although it cannot be ruled out that the uncharacterized solids could be derivatives of these. However, one clear result was the isolation of a liquid with boiling point 40°C. described as an air-oxidizable base that formed salts with acids. Thenard reported the elemental analysis of this compound to be: %C, 46.1 and 46.08, and %H, 11.6 and 11.7. These data do support the formula $(CH_3)_3P$ (calcd. %C 47.4, %H 11.8), written by Thenard as C^6H^9P which he derived from the atomic weight of C to be 6, not 12. In later years Augustus Cahours and

August Hofmann [10] determined the boiling point of authentic trimethylphosphine to be 40-42°C. In modern times, a similar value (37.8° C.) has been reported [16]. Another liquid had b.p 250° C., and was spontaneously flammable on exposure to air. Thenard stated that it had the (empirical) formula Me_2P. It is possible that this compound is tetramethylbiphosphine, Me_2P-PMe_2, suggested also by Hofmann [10], but this compound was reported many years later to have a much lower b.p. of 140.2 °C. [17].

From this analysis of Thenard's work, statements like "Thenard --prepared several phosphines—" [1a] and "Thenard prepared a series of phosphine derivatives---" [18] need clarification, since there is no evidence to support this. References still appear [19] that suggest he might have accomplished this. Indeed, it was not until several years after Thenard's publications that the concept of substituting alkyl groups for the hydrogen of ammonia, thus to give primary, secondary and tertiary amines, was developed [9]. Nevertheless, Thenard's pioneering work clearly pointed to the fact that phosphorus compounds with carbon could exist and he was the first to make this discovery. He made the interesting speculation that a great number of new compounds might be possible, and even that phosphorus compounds might occur in cerebral matter and nerves (why he thought this is not clear); the former was indeed prophetic, but no carbon-phosphorus compounds are known to be biosynthesized in humans and other mammals, although they are common in lower life forms, especially invertebrate marine animals [20].

The next publication on an organic derivative of phosphorus was made by F. Berle at Wiesbaden [21] in 1855. He used a process related to that of Thenard; he first prepared pure sodium phosphide from sodium and white phosphorus, and then performed an alkylation reaction with ethyl iodide. In a stream of carbon dioxide, he distilled out a liquid (yield and b.p. not given) that he proposed to have the formula $(C_4H_5)_3P$. As noted, at that time the atomic weight of carbon was taken as 6 rather than 12, a practice that lasted up to about the 1850's, so his formula is that of triethylphosphine (indeed, which he called "phosphortriäthyl"). Berle heated this product with ethyl iodide and obtained a difficult to purify crystalline solid (melting point not given), which had an iodine content of 45.5% and 47.8% for different samples. He wrote the formula $(C_4H_5)_4P$, I (calcd. for I, 46.3%) and his paper can be taken as confirmation of Thenard's results that alkylated phosphorus can exist and thus that there can be a large, totally unexplored field of organophosphorus compounds. We might note that Berle did not observe any other P-containing products such as those mentioned by Thenard, which might be associated with Berle's use of a purer form of a metallic phosphide.

CONCLUSION

By 1857, it had been observed and confirmed that compounds with carbon-phosphorus bonds could exist, at least in the form of tertiary phosphines (trimethyl and triethyl). The way was opened for wide exploration of this new field, which commenced with pioneering work by August Hofmann and August Michaelis, and eventually many others during the nineteenth century.

REFERENCES

[1] Kosolapoff, G. M. *Organophosphorus Compounds*; John Wiley: New York, 1950; (a) p vi, (b) p 2, (c) pp 25-26.
[2] Quin, L. D. *Heteroatom Chem.* 2013, *24*, 243.
[3] Lassaigne, J.-P. *Annales chim.* 1820, *13*, 274.
[4] Pelouze, J. *Ann. pharm.* 1833, *6*, 129; Pelouze, J. *Annales chim.* 1833, *52*, 37. See also Liebig, J. *Ann. pharm.* 1833, 6, 149.
[5] Russell, C.A. In *Instruments and Experimentation in the History of Chemistry;* Editors Holmes, F.L.; Levere, T.H.; MIT Press: Cambridge, MA, 1999; pp 312-314.
[6] Rigaudy, J. P.; Klesney, S. P. *Nomenclature of Organic Compounds*; Pergamon Press: Oxford, UK, 1979; pp 382-408.
[7] Rocke, A. J. *The Quiet Revolution: Hermann Kolbe and the Science of Organic Chemistry;* University of California Press: Berkeley, 1993.
[8] Rocke, A. J. *Image and Reality*; University of Chicago Press: Chicago, 2010.
[9] Hofmann, A. W. *Phil. Transactions* 1850, *93*; 1851, 357.
[10] Cahours, A.; Hofmann, A. W. *Ann.* 1857, 104, 1. Repeated in English: Hofmann, A.; Cahours, A. *Quart. J. Chem. Soc.,* 1859, 11, 56. A preliminary note appeared in 1855: Cahours, A; Hofmann, A. W. *Compt. rend.* 1855, *41*, 831.
[11] Crum Brown, A. *Trans. Roy. Soc. Edinburgh* 1864, *23*, 707. [12] Ref. 8, p. 155.
[13] Michaelis, A. *Ann.* 1876, *181*, 265.
[14] Ref. 7, p. 298.

[15] Thenard, P. *Compt. rend.* 1845, *21*, 144; *Compt. Rend.* 1847, *25*, 892. For a review of this work in English, see *Philosophical Magazine* 1848, *32*, 153.
[16] Rosenbaum, E. J.; Sandberg, C. R. *J. Am. Chem. Soc.* 1940, *62*, 1622.
[17] Burg, A. *J. Am. Chem. Soc.* 1961, *83*, 2226.
[18] Hartley, F. R. *The Chemistry of Organophosphorus Compounds*; John Wiley: Chichester, U. K.,1990; p 2.
[19] Corbridge, D. E. C. *Phosphorus 2000, Biochemistry and Technology;* Elsevier: Oxford, 2000; pp 3-4.
[20] Quin, L. D. *A Guide to Organophosphorus Chemistry*; John Wiley: New York, 2000; pp 357-362.
[21] Berle, F. *J. Prakt. Chem.*1855, *66,* 73; *Ann.* 1856, *97*, 334.

Chapter 2

THE PIONEERING WORK OF AUGUST WILHELM HOFMANN IN ORGANOPHOSPHORUS CHEMISTRY*

ABSTRACT

August Wilhelm Hofmann, 1818-1892, was the first chemist to conduct extensive experimentation with organic phosphorus compounds, and he discovered many of the functional groups based on phosphorus, including primary, secondary and tertiary phosphines, oxides and sulfides of the latter, phosphonic and phosphinic acids and their chloro derivatives, dithiophosphinic acids, tetraalkyl phosphonium hydroxides, Lewis salts from both tertiary phosphines and their oxides. Some of his procedures are still in use today. Some fundamental properties of these groups were also discovered. August W. Hofmann justly deserves recognition as the originator of the field of Organophosphorus Chemistry

2.1. INTRODUCTION

It was the pioneering work of August Wilhelm Hofmann beginning in 1855 that revealed the existence of a new and large arena of stable compounds with carbon-phosphorus bonds, in many types of novel functional groups. His many significant observations, still of importance today, are reviewed in this chapter. The argument has been made [1] that he should be considered the originator, even the Founding Father, of the field of Organophosphorus Chemistry. His stature and contributions were so important that it is appropriate for this account of his work to be preceded by a brief biographical outline of his truly outstanding career.

2.2. A Brief Biography of August Wilhelm Hofmann (Later von Hofmann), 1818-1892

Figure 2.1. A. W. Hofmann at the time of his position at the Royal College of Chemistry. Reprinted from Playfair, L. *J. Chem. Soc. Trans.* 1896, 69, 575.

There are several biographies available on August Hofmann; many are listed in items on the internet, especially in Wikipedia and Encyclopedia.com. The most informative is found in a series of memorial lectures published in 1896, a few years after his death, in the *Journal of the Chemical Society Transactions* [2]. Of special relevance to the present review is the section by Henry E. Armstrong where on pp. 671-684 he summarizes some of Hofmann's work on phosphorus compounds. A largely non-technical account of

Hofmann's career has been published by Christof Meinel and Hartmut Scholz in the *International Edition of Angewandte Chemie* [3]. This article also appeared in the German book by these authors *"Die Allianz von Wissenschaft und Industrie"* [4]. Also included in this book is a chapter (page 299- 316) reviewing Hofmann's work on phosphorus chemistry, by Herbert Teichmann. The great originality and especially the relevance today of much of Hofmann's work are made clear in this chapter.

Only some essential biographical details for August Hofmann are given here. He was born in Giessen, Germany in 1818, and died in Berlin in 1892. He was a student of the famous Justus Liebig and was awarded the doctoral degree by the University of Giessen in 1841. He later was appointed to a position at the University of Bonn. In 1845 Hofmann became Director of the new Royal College of Chemistry in London; he remained in London until he returned to the University of Bonn in 1864 and then shortly thereafter accepted the Chair of Chemistry at the University of Berlin (1865). As will be seen, Hofmann's early work on phosphorus compounds was performed during his period in London; this work was continued in Berlin. He remained in Berlin for the rest of his life. Hofmann was one of the most famous and accomplished of the great group of organic chemists of the nineteenth century. He made major contributions to the development of the coal tar industry and later to the new dye industry in association with his former student, the well-known William Henry Perkin. He is primarily remembered by organic chemists today for his fundamental studies on nitrogen chemistry (*vide infra)* and as the founder of the Deutsche Chemische Gesellschaft and its famous chemical journal, the "Berichte“. He also was the first to synthesize many common organic chemicals. However his pioneering work opened up the new field of Organic Chemistry where carbon is directly bonded to phosphorus. Surprisingly, as pointed out in reference [1], more recognition is given to the extensive later work of Carl Arnold August Michaelis (the subject of Chapter 3). Thus in the highly valuable book by G. M. Kosolapoff [5a,b], the first in English on organophosphorus chemistry, only Michaelis and A. E. Arbuzov are mentioned as the originators of the field. A second edition with L. Maier [6] does state in the introduction that Hofmann was the first to synthesize alkylphosphonic acids. While this is true, he did far more than that. Similar statements can be found in other major works on phosphorus chemistry. Thus in his series of volumes entitled "*The Chemistry of Organophosphorus Compounds* [7], F. R. Hartley notes the significant 1845 paper [8] of P. Thenard (which is discussed in Chapter 1) as the starting point in this field, from which "progress was fairly rapid. Michaelis ---- dominated the field for many years--". Hofmann is not mentioned. D. E. C. Corbridge [9] also mentions the paper of Thenard and goes on to say "The most notable pioneers in this field were Michaelis (1847-1916) and Arbusov (1877-1968) who are now regarded by many as the founders of organophosphorus chemistry." No doubt this statement is true, but Hofmann was there first and again is not mentioned.

Hofmann's famous long-term emphasis on nitrogen chemistry started with his pioneering work on the bases of coal tar. As noted, it was he who first described the primary, secondary and tertiary amines with various alkyl substituents, which also played a role in the development of the Structural Theory propounded by Kekulé, Butlerov, Erlenmeyer, Crum Brown and others in the 1850's and 1860's and which revolutionized organic chemistry. His name is still prominent today in the discipline of Organic Chemistry, and all students are familiar with the Hofmann degradation of amides to amines, with the Hofmann elimination of amines from quaternary hydroxides to form alkenes but of different regiochemistry from those of elimination of halides (the Hofmann Rule), with the Hofmann exhaustive methylation procedure for the determination of the structure of natural amines.

It is relevant to consider the research time lines of the three pioneering chemists studying organophosphorus compounds, Hofmann, Michaelis and Arbuzov. Hofmann published all of his reports (except one) in the period 1855 through the first half of 1873. The exception was a clarification in 1892, the year of his death, of work published in 1871. Michaelis's first papers on phosphorus chemistry, of which four have been found [10], were published in 1872. However, these focused on the inorganic compounds $PSBr_3$ and PCl_3Br_2 and included only in part the reaction with alcohols and mercaptans to give phosphoric acid esters. His far more important publications on organophosphorus compounds where phosphorus is bonded to carbon began in late 1873 [11], thus after the period of Hofmann's work. He continued to publish on such compounds until 1915. Therefore there was no overlap with Hofmann on publications in this field. The first paper by A. E. Arbuzov on phosphorus compounds appeared in 1905, and he continued research in this field until 1957. The research of these three famous chemists spanned a continuous period of 100 years. While they dominated the field through these years, it will be seen that Hofmann's work on C-P compounds was of great breadth and originality in its time, and was performed before Michaelis became active.

In this Chapter, prepared from a study of the original German literature, Hofmann's numerous significant accomplishments that opened this difficult field will be described and evaluated. It will be seen that the argument [1] that he deserves recognition as the originator, even as Founding Father, of the field of Organophosphorus Chemistry, is well substantiated. This presentation here of his great research accomplishments is derived from a published summary of his work by L.D. Quin [1].

2.3. Research Accomplishments of August W. Hofmann in Organophosphorus Chemistry

Hofmann's research in phosphorus chemistry started with a study of the alkyl derivatives of phosphine, following his discoveries in amine chemistry, and he was the first to synthesize and characterize a family of primary, secondary and tertiary phosphines and describe them as such. In these studies, he discovered many of the other functional groups containing phosphorus that we are so familiar with today, and which are the basis for the formation of countless compounds.

Hofmann working with Augustus Cahours, an accomplished French chemist, recognized the significance of Thenard's research on phosphines [8], and the possible analogy of phosphines to his own amine chemistry. With the progress made in the ten years after his publications on amines [12], including advances in theory as outlined in Chapter 1 (recognition of the concept of group substitution, and the rise of the Structural Theory), they examined again the products of the reaction of an alkyl halide with a metallic phosphide. They chose to use methyl iodide and sodium phosphide (Na_3P) in these studies, and without giving details (this is typical of the times; full experimental details and even yields are frequently not stated) reported that three compounds were obtained [13]. This is the same process described independently and at about the same time by Berle [14]. The three products were: a liquid with empirical formula Me_2P (the actual formula used by Hofmann, and said to resemble the arsenic compound "Kacodyl" (now written as Me_2 As-$AsMe_2$) and possibly Me_2P-PMe_2 but unconfirmed; a second liquid with formula Me_3P (consistent with the compound obtained by Thenard); a crystalline solid recognized to be a quaternary salt with formula Me_4PI (suggesting that possibly one of the solids observed by Thenard was Me_4PCl). This reaction was a difficult and dangerous process, and to replace it a new way for obtaining phosphines was invented: the displacement of the three chlorines of "terchloride of phosphorus", PCl_3, by reaction with zinc alkyls [13]. This was expressed by

Hofmann by the quite modern equation (1), except that he wrote this shortly before the reform of atomic weights took place. Thus his "zinc -methyl", written as MeZn, is properly Me_2Zn, and zinc chloride is $ZnCl_2$. The zinc alkyls had been introduced by Edward Frankland in his pioneering studies around 1850.

$$PCl_3 + 3MeZn = 3ZnCl + Me_3P \quad (1)$$

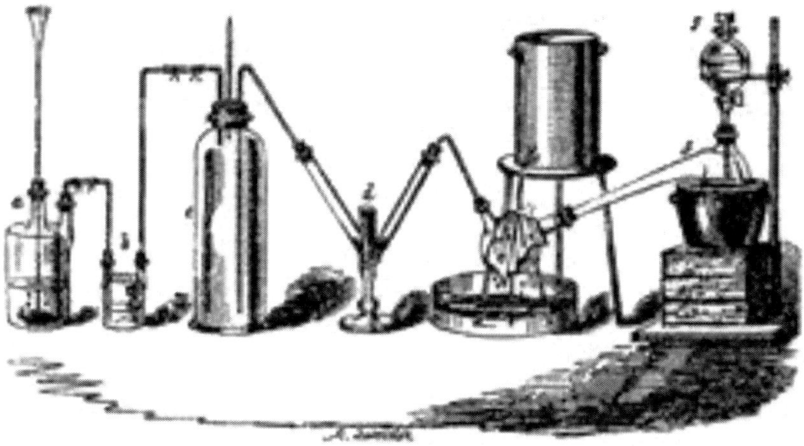

Figure 2.2. Apparatus used by Hofmann for the reaction of zinc alkyls with phosphorus trichloride. Reprinted from Cahours, A.; Hofmann, A. *Ann.* 1857, *104*, 1.

Hofmann found it easier to work with the ethyl series than methyl, and most of his work was conducted with triethylphosphine. He named the products as we would today, thus as trialkylphosphines. He found that the phosphines were bound to "ZnCl", and this is the first demonstration of the profoundly important property of the phosphines acting as Lewis bases (and later as ligands in coordination complexes). To release the free phosphine, he reacted the zinc chloride complex with alkali. Hofmann also pointed out that the trialkylphosphines were bases of strength comparable to that of his trialkylamines. It hardly needs to be mentioned that the reaction of PCl_3 with organometallics is the major route to symmetrical trialkylphosphines used even today. Hofmann remarked on how dangerous the process is, with inflammations, explosions, horrible smells, and products with toxic reaction to humans (headaches and sleeplessness). It took great courage to persist in this work, no doubt conducted in the open laboratory without adequate hoods or ventilation, and likely without the equivalence of modern safety devices, especially without plastic shields or safety glasses. To simplify the process, Hofmann designed a special apparatus shown as Fig. 2.2. The part of the apparatus on the left is for generation of CO_2 to keep the process free of oxygen, and thus he is an early user of the common modern technique of blanketing air-sensitive materials with an inert gas. The reaction itself is conducted in the retort on the right by dropping the PCl_3 into diethylzinc contained therein. The "V" in the center of the apparatus is a trap to prevent back-up of materials.

Hofmann made many original discoveries during this work with triethylphosphine, which revealed its rich chemistry, and provided background information still of value today.

- He established its composition by elemental analysis, and assigned its formula (**1**) shown in the style of the times (and still with the atomic weight of C being 6) and expressing ethyl as E, to be:

$$C_{12}H_{15}P = \left.\begin{matrix} E \\ E \\ E \end{matrix}\right\} P$$

1

- He discovered that it is a base, with strength resembling his triethylamine, and that it forms crystalline salts with HCl, HBr, HI, HNO_3, and H_2SO_4. He did little work with the salts because they were extremely hygroscopic.
- He found that triethylphosphine reacted rapidly with oxygen of the atmosphere to form a crystalline compound. This was the first observation of the now vast family of phosphine oxides. He expressed its formula as E_3PO_2 and referred to it as a "binoxide" because of the contemporary notion that oxygen had an atomic weight of 8. He later wrote it correctly as E_3PO.
- He prepared the first example of a phosphine sulfide by reacting elemental sulfur with triethylphosphine, assigning the structure E_3PS_2 (derived from the atomic weight of sulfur being 16, not 32, and today written Et_3PS). Similarly, he prepared the first example of a phosphine selenide (his E_3PSe_2).
- He found that platinum chloride ($PtCl_2$) formed a crystalline product with triethylphosphine, whose formula he expressed as

$C_{12}H_{15}P,HCl,PtCl_2$, and thus introduced the still-practiced procedure of characterizing oxidizable phosphines as stable solid derivatives.

- He formed the crystalline salt tetraethylphosphonium iodide (the actual name he correctly assigned) with ethyl iodide, recognizing the process as analogous to the formation of quaternary ammonium iodide salts formed from trialkylamines. Berle [14] as noted in Chapter 1 reported this compound at about the same time.
- Hofmann had previously prepared quaternary ammonium hydroxides by reacting quaternary iodides with silver oxide and water. He applied this procedure to his tetraethylphosphonium iodide and obtained exactly the same result, thus being the first to report a tetraalkylphosphonium hydroxide. The product was extremely deliquescent and was a base as strong as KOH. It rapidly took up carbon dioxide from the atmosphere. It was isolated as a crystalline solid but its properties defied precise elemental analysis. Hofmann expressed its composition as $C_{16}H_{21}PO_2$ and E_4PO,HO, again using atomic weights C=6 and O=8. The correct formulation would be Et_4POH, or in modern ionic form $Et_4P^+ OH^-$

Hofmann found a very significant difference in the behavior of quaternary ammonium and phosphonium hydroxides. The former he knew on heating to decompose, expressed as he wrote it by Eq. 2.

$$E_4NO,HO = C_4H_4 + 2HO + E_3N \qquad (2)$$

His C_4H_4 corresponds to our C_2H_4, which he correctly identified as ethylene. This is the basis for the well-known Hofmann elimination used in determining the structure of amines. The process has been well studied in the recent years of mechanistic organic chemistry because if the alkene is capable of existing in isomeric forms, that which is favored has the smallest number of carbon substituents on the double bond (the Hofmann Rule), while that from HX elimination from alkyl halides with base has the greater number of carbon substituents (the Saytzeff Rule). When the process was applied to quaternary phosphonium hydroxides (which he called "hydrated oxide of tetraalkylphosphonium"), yet another pathway was followed, the eliminated hydrocarbon being a saturated alkane. Hofmann recognized the phosphorus product from tetraethylphosphonium hydroxide to be the same as the one he obtained from air-oxidation of triethylphosphine, thus triethylphosphine oxide. Here is the equation (3) he wrote to express this chemistry:

$$E_4PO,OH = E_3PO_2 + EH \quad (3)$$

Today we would express his reaction as Eq. 4.

$$Et_4P^{+}\ OH^{-} \longrightarrow Et_3P{=}O \ + \ CH_3CH_3 \quad (4)$$

Hofmann went on to prepare trimethylphosphine by the PCl_3-Zn alkyl route, finding it to have similar properties to those of triethylphosphine. He also prepared quaternary salts from trimethylphosphine with ethyl iodide and n-amyl iodide.

To summarize this remarkable paper describing so much original and difficult experimentation, we can say that Hofmann opened the field of organophosphorus chemistry by demonstrating for the first time the existence of several P-containing functional groups and of some reactions inter-relating them, these groups being *trialkylphosphines, trialkylphosphine oxides, trialkylphosphine sulfides, trialkylphosphine selenides, Lewis salts of trialkyl phosphines (with zinc chloride and platinum chloride), tetraalkylphosphonium halides and tetraalkylphosphonium hydroxides.*

In another paper published two years later [15] Hofmann reported improved procedures for preparing and purifying triethylphosphine and its oxide. By this time he was showing the correct formula as $(C_2H_5)_3PO$. He also reported a new observation, that phosphine oxides also form Lewis salts with certain metallic compounds just as phosphines do. Mostly he studied the salts from triethylphosphine oxide and zinc iodide and platinum chloride ($PtCl_2$). Further studies on phosphine reactions with sulfur and sulfur compounds were reported. Probably the most interesting was the reaction of tertiary phosphines with carbon bisulfide, whereupon red crystals were formed. He presented the molecular formula we now know to be correct, R_3P-CS_2. This product has intrigued chemists for many years (e. g., see [5c]). Its structure may be expressed as **2**.

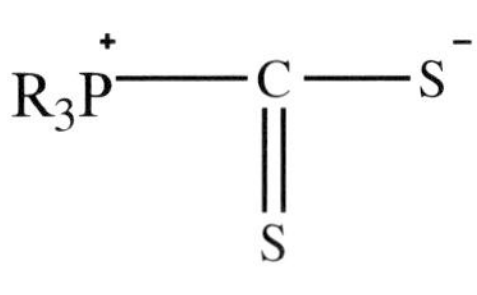

2

Studies were also conducted on reaction products with "sulfocyanate of phenyl" (presumably C_6H_5-N=C=S) and "cyanate of phenyl" (meaning the isocyanate, C_6H_5-N=C=O). The latter study is of significance because in it Hofmann performed the first organic reaction with a phosphine acting as a catalyst, a type of process of great importance today. Hofmann reported that dipping a stirring rod into triethylphosphine and then into a body of (liquid) "cyanate of phenyl" (actually isocyanate; Hofmann sometimes did not distinguish between the two) produced a strong exothermic reaction and the solidification of the liquid. The product was reported to be "cyanurate of phenyl", a trimer of the cyanate. A similar result took place with "cyanate of ethyl". However, Hofmann later realized that the product was a dimer, because it reacted with alcohols to form esters of diphenylallophanic acid (**4**, a ureide of carbonic acid) [16]. He referred to the dimer as "dicyanate"; it is indeed a diazacyclobutanedione, whose structure (**3**) was much later confirmed by X-ray crystallography [17].

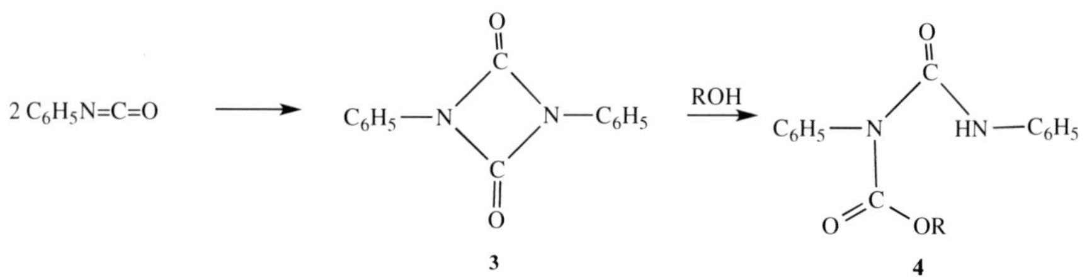

(5)

Triethylphosphine was said to catalyze the trimerization of isocyanic acid to form cyanuric acid. Cyanuric acid was a well-known compound at the time, first characterized by Scheele in 1776 as a trimer of cyanic acid.

In later papers [18], Hofmann reported the first synthesis of a phosphorus compound containing a carbon functional group, the forerunner of a vast field developing over the years. He reacted triethylphosphine with ethyl chloroacetate to obtain a phosphonium salt with a $-CH_2COOEt$ substituent. As formed, the product would be Et_3P^+-CH_2COOEt Cl^-, and Hofmann characterized it as such by forming a complex salt with platinum chloride ($PtCl_2$). The ethyl group and chloride of the ester were removed by the action of moist silver oxide to give a compound we would now represent with the dipolar formula $Et_3P^+CH_2COO^-$ which is a type described as a betaine. Hofmann recognized it as a phosphorus relative of the nitrogen-containing family derived from glycocoll, the name in use then for glycine (NH_2CH_2COOH). The name betaine is used for the parent structure, which would be Me_3N^+-CH_2COO^- and which had been isolated from sugar beets. Hofmann's work inspired A. H. Meyer to use his procedure to prepare the phosphorus counterpart (Me_3P^+-CH_2COO^-) of betaine [19]. This is the first phosphorus compound to be modeled after a natural product.

Hofmann later developed a new synthesis of tetraalkylphosphonium iodides, involving the reaction of an alcohol with phosphonium iodide (PH_4I) at 160-180°C. in a sealed tube [20]. Presumably an alkyl iodide was first formed which then alkylated phosphorus. By this method he synthesized Et_4PI and Me_4PI as crystalline solids giving the correct iodine analysis. He later used methyl and ethyl iodides directly in the reaction with PH_4I [21], and then in important modifications of the ratio of reactants, and in the presence of zinc oxide as an acid acceptor, he was able to accomplish the first synthesis of a primary ($EtPH_2$) and a secondary (Et_2PH) phosphine. He clearly recognized the analogy with amine chemistry [22]. He went on to prepare both methylphosphine and dimethylphosphine by this method [23], and two years later he reported the synthesis of the various phosphines in the propyl, butyl, and amyl series [24].

Hofmann was the first chemist to synthesize a member of the highly important family of phosphorus acids bearing a carbon-phosphorus bond [25]. He passed a stream of the gaseous methylphosphine into boiling nitric acid as an oxidant, and was able to isolate a solid product, m.p. 105° (confirmed in modern work), which he recognized to be a dibasic acid. Elemental analysis allowed him to write the formula, $(CH_3)H_2PO_3$ for the solid, and the formula $(CH_3)M_2PO_3$ for salts that he prepared from several metallic compounds. Hofmann assigned the name methylphosphinic acid to his new acid; today it is called methylphosphonic acid, with the name phosphinic acid reserved for acids with two carbon to phosphorus bonds. We would write the formula in the remarkably close form of $CH_3P(O)(OH)_2$. Hofmann went on to oxidize dimethyphosphine, obtaining the acid (in his form, $(CH_3)_2HPO_2$) which he also converted to various salts. He gave this new acid the name dimethylphosphinic acid, which is the name we use today. He proceeded to make the corresponding acids in the ethyl series by the same reactions applied to the ethyl phosphines. Hofmann clearly recognized the fact that his new acids fit into a series from the replacement of OH groups on phosphoric acid, and he wrote the series seen in Scheme 1.

"Orthophosphorsäure $\left.\begin{matrix} HO \\ HO \\ HO \end{matrix}\right\} PO$

"Methylphosphinsäure" -- $\left.\begin{matrix} CH_3 \\ HO \\ HO \end{matrix}\right\} PO$

"Dimethylphosphinsäure"-- $\left.\begin{matrix} CH_3 \\ CH_3 \\ HO \end{matrix}\right\} PO$

"Trimethyl phosphineoxyd" -- $\left.\begin{matrix} CH_3 \\ CH_3 \\ CH_3 \end{matrix}\right\} PO$

Scheme 1. Hofmann's phosphoric acid substitution sequence.

Note that these formulas are getting close to true structural formulas, but without bonds, or any concept of 3-dimensional structure. Hofmann also recognized the similarity to the methyl-replacement series based on arsenic, starting with arsenic acid, $As(O)(OH)_3$. In further work [26] Hofmann prepared several other alkylphosphonic and dialkylphosphinic acids. In this paper, he also reported the first synthesis of a phosphonic dichloride, $CH_3P(O)Cl_2$, finding m.p. 32°C. and b.p. 163°C., values well known today. His method involved reaction of methylphosphonic acid with PCl_5, just as we might do today. He also reported the first dialkylphosphinic chloride, prepared in the same way.

Attempts to prepare phenylphosphine by his PH_4I alkylation method as applied to chlorobenzene were failures, as we would expect today from knowledge of the relative inertness of aryl halides to nucleophilic substitution. However, Hofmann did succeed in making benzylphosphine, the first phosphorus compound to contain an aryl group [27]. He reported another product as dibenzylphosphine, but it was suggested many years later that this product was actually tribenzylphosphine oxide [28].

From a complicated series of experiments on the reaction of ethylene dibromide with triethylphosphine [29], three results may be worth noting here. (1) A bis(phosphonium) salt of the formula **5** was obtained and was the first of this type:

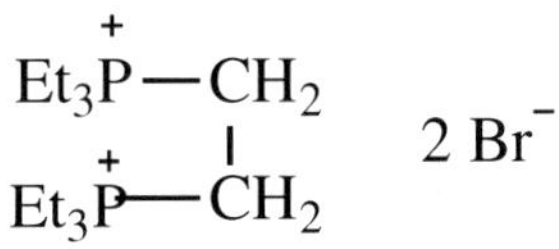

5

(2) The first phosphorus compound with a vinyl group, $CH_2{=}CHPEt_3^+ Br^-$ was reported, (3) As with halo benzenes, vinyl halides did not form phosphonium salts with triethylphosphine, for the same reason.

In his studies on secondary phosphines [22], Hofmann observed that a reaction occurred with elemental sulfur to form an oily, non-crystallizing compound, but did no further work on it at that time. He actually had prepared the first example of a dialkyldithiophosphinic acid, yet another of the numerous functional groups of phosphorus that he discovered. Many years later with F. Mahla [30] he re-examined this reaction as applied to diethylphosphine and was able to obtain a crystalline ammonium salt with m.p. 193°C., as well as a silver salt. The free acid was isolated as an oil, which gave a reasonably good elemental analysis. Hofmann properly named the new compound diethyldithiophosphinic acid. In 1946, Malatesta and Pizzotti [31] repeated this work successfully and further characterized Hofmann's compound as metallic derivatives. Thus we can add the creation of a representative of the family $R_2P(S)(SH)$ to the list of Hofmann's accomplishments in the field of organophosphorus chemistry. Hofmann had returned to this problem after a gap of some 21 years, during which he reported no further work with phosphorus compounds, odd after his burst of activity in 1871-1873 in this field. The paper appeared shortly after his death. Hofmann also reported the benzoyl derivative of diethyldithiophosphinic acid which he prepared by reaction of the sodium salt with benzoyl chloride and wrote the structure $(C_2H_5)_2PS.SCOC_6H_5$. This compound had m. p. 54°C. and was characterized by elemental analysis; it too is the first of its type.

CONCLUSION

August Hofmann thus opened a wide and important field of Organic Chemistry when he extended his research on nitrogen chemistry to a study of compounds of phosphorus. Even though his work was being done during a time of continual revision of theories on the constitution of organic chemicals, he was able to understand and interpret the chemical changes he observed in remarkably clear ways, and to use some formulas that are very similar to those we use today. During his research, he discovered many of the common functional groups that are at the basis of organophosphorus chemistry, as is detailed in this chapter. He showed great chemical insight in interpreting his laboratory results. Laboratory work with some classes of phosphorus compounds can be very difficult and dangerous, but these problems did not deter him.

In the memorial to Hofmann published in 1896 [2], Armstrong makes the following observation: "Although in the interval since his researches were published our knowledge of the phosphines and allied compounds has been considerably increased, chiefly owing to the discovery of phenyl derivatives by Michaelis and his pupils, in the case of this group, as in that of the amines and substituted ammonium compounds, we owe all that is fundamental to Hofmann, the properties which he has shown to be characteristic of the primary, secondary and tertiary ethyl derivatives of phosphine being characteristic of all other similarly constituted compounds."

In this chapter, Hofmann's work has been reviewed from a modern perspective, some 115 years after Armstrong's observation on phosphines. It is obvious that Hofmann's research led to several other new functional groups, work on which has become of enormous scope and importance over the years. He deserves the recognition as the Founding Father of Organophosphorus Chemistry as proposed in reference [1].

References

[1] Quin, L. D. *Heteroatom Chem.* 2013, *24*, 243.
[2] Playfair, L. *J. Chem. Soc. Trans.* 1896, *69*, 575.
[3] Meinel, C.; Scholz, H. *Angew. Chem. Int. Ed. Engl.* 1992, *31*, 1265.
[4] Meinel, C.; Scholz, H. *Die Allianz von Wissenschaft und Industrie: August Wilhelm Hofmann, 1818-1892, Zeit. Werk. Wirkung*; VCH: Weinheim, Germany, 1992.
[5] Kosolapoff, G. M. *Organophosphorus Compounds*; John Wiley: NY, 1950; (a) p vi, (b) p 2, (c) p 25-26.
[6] Kosolapoff, G. M.; Maier, L. *Organic Phosphorus Compounds*, Vol. 1; John Wiley: NY, 1972; p ix.
[7] Hartley, F. R. *The Chemistry of Organophosphorus Compounds*; John Wiley: Chichester, U. K., 1990; p 2.
[8] Thenard, P. *Compt. rend.* 1845, *21*,144; *Compt. rend.* 1847, *25*, 892.
[9] Corbridge, D. E. C. *Phosphorus 2000, Biochemistry and Technology*; Elsevier: Oxford, 2000; pp 3-4.
[10] Michaelis, A. *Ber.* 1872, *5*, 4, 6, 9; Michaelis, A. *Ann.* 1872, *164,* 9.
[11] Michaelis, A. *Ber.* 1873, *6,* 601, 816.
[12] Hofmann, A. W. *Phil. Transactions* 1850, *93*; 1851, 357.
[13] Cahours, A.; Hofmann, A. W. *Ann.* 1857, 104, 1. Repeated in English: Hofmann, A.; Cahours, A. *Quart. J. Chem. Soc.,* 1859, 11, 56. A preliminary note appeared in 1855: Cahours, A; Hofmann, A. W. *Compt. rend.* 1855, *41*, 831.
[14] Berle, F. *J. Prakt. Chem.* 1855, *66*, 73; *Ann.* 1856, *97*, 334.
[15] Hofmann, A. W. *Quart. J. Chem.Soc.* 1861, *13*, 289.
[16] Hofmann, A. W. *Ber.* 1871, *4,* 246.
[17] Millar, A. T.; Springall, H. D. *A Shorter Sidgwick's Organic Chemistry of Nitrogen*; Clarendon Press: Oxford, 1969; pp 362-363.
[18] Hofmann, A. W. *Proc. Roy. Soc.* 1862, *11*, 25; *Compt. rend.* 1862, *54,* 252.
[19] Meyer, A. H. *Ber.* 1871, *4*, 734.
[20] Hofmann, A. W. *Ber.* 1871, *4*, 205.
[21] Hofmann, A. W. *Ber.* 1871, *4*, 372.
[22] Hofmann, A. W. *Ber.* 1871, *4*, 430.
[23] Hofmann, A. W. *Ber.* 1871, *4,* 605.
[24] Hofmann, A. W. *Ber.* 1873, *6*, 292
[25] Hofmann, A. W. *Ber.* 1872, *5,* 104.
[26] Hofmann, A. W. *Ber.* 1873, *6*, 303.

[27] Hofmann, A. W. *Ber.* 1872, *5,* 100.
[28] Letts, E. A.; Blake, R. N. *Trans. Roy. Soc. Edinburgh* 1889, *35,* 527.
[29] Hofmann, A. W. *Quart. J. Chem. Soc.* 1861,*14,* 73.
[30] Hofmann, A. W.; Mahla, F. *Ber.* 1892, *25,* 2436.
[31] Malatesta, L.; Pizzotti, R. *Gazz. Chim.. ital.* 1946, *76,*167.

Chapter 3

THE MANY RESEARCH ACCOMPLISHMENTS OF AUGUST MICHAELIS

ABSTRACT

August Michaelis dominated the field of organophosphorus chemistry from 1873 until his death in 1916. He was the first to make compounds with phosphorus bonded to the benzene ring. He prepared many such compounds with phosphorus present in various functional groups. He also prepared compounds with phosphorus bonded to alkyl groups, and he devised two synthetic methods which still have great value in the synthesis of phosphorus acid derivatives. These are known today as the Michaelis-Becker and the Michaelis-Arbuzov Reactions.

3.1. INTRODUCTION

There can be no doubt that all practicing organophosphorus chemists are familiar with the name August Michaelis, since it is perpetuated in connection with still-important processes: the Michaelis-Becker synthesis of alkylphosphonic acids based on alkylation of anions from dialkyl hydrogenphosphonates ($HP(O)(OR)_2$), and the Michaelis-Arbuzov synthesis also of phosphorus acids from alkylation of tricovalent phosphorus esters (most commonly of trialkyl phosphites). But Michaelis's accomplishments extended far beyond that; he synthesized many new classes of organophosphorus compounds and described their properties, providing a body of information that we still depend on today. In this Chapter his major discoveries will be described. It must be remembered that classes of compounds with phosphorus in the trivalent state, are very difficult to work with, many having terrible smells, toxicity, and may have high reactivity to oxygen. These properties point all the more to the outstanding early work of Michaelis and his students.

It is important to remember his period of research in the early chronology of the development of organophosphorus chemistry, as discussed in Chapters 1 and 2. The very first discovery of a compound with a C-P bond had been made by the French chemist Paul Thenard in 1845. August Hofmann's extensive work was published in the period 1855 to 1873 (with one late report in 1892). Michaelis's first work with phosphorus compounds was published in 1872 and was only concerned with derivatives of phosphoric acid [1]. He entered the field of phosphorus compounds with bonds to carbon in 1873 [2] and he continued this work until his death in 1916. A.E. Arbuzov, the third great name in organophosphorus chemistry, first published on his work in 1905 and was active until 1948. His contributions thus appeared in the twentieth century and are not included in this book. In this Chapter, the compounds studied by Michaelis are drawn in structural detail, which is seldom found in his original publications. It is felt this makes the chemistry more readily understandable.

3.2. A Brief Biography of K. A. August Michaelis, 1847-1916

Karl Arnold August Michaelis, later using only August Michaelis as the author of his publications, was born in Germany in Bierbergen, Hannover, December 26, 1847. He studied at Göttingen and Jena Universities. He received an appointment at the Technischen Hochschule of Karlsruhe, where he became Professor in 1876. In 1880, he took up a new position as Professor at the University of Aachen, and then made his final move in 1890 to the University of Rostock as Professor, where he remained until his death on January 31, 1916.

Michaelis was a pioneer in two related fields, the organic chemistry of phosphorus and also of arsenic. According to a short obituary published immediately after his death [3], it was his work on arsenic compounds that led Paul Ehrlich to his monumental discovery of the chemotherapeutic activity of such compounds, notably of Salvarsan for the treatment of syphilis. Michaelis's research on phosphorus compounds spanned his entire academic career. As noted, his first paper on carbon-phosphorus compounds was published in 1873 [2a] and was of major importance as it opened up access to many unknown compounds containing P-aryl functions. His outstanding research accomplishments over the years established him internationally as the preeminent organophosphorus chemist of his day.

Figure 3.1. K. A. August Michaelis in his Rostock years.

It is convenient to divide his work into the most significant sub-topics:

1. The first synthesis of arylphosphonous dihalides ($ArPX_2$, earlier known as aryl dihalophosphines), giving access for the first time to a host of P-aryl compounds.
2. The synthesis and characterization of P-arylphosphines, most significantly triphenylphosphine, which is even today the most important of all phosphines.
3. The first synthesis of arylphosphonic acids ($ArP(O)(OH)_2$).
4. The synthesis of alkylphosphonic acids based on the establishment of the P-C bond by the reaction of alkyl halides with sodium derivatives of dialkyl esters of phosphorous acid, $HP(O)(OR)_2$ (known today as the Michaelis-Becker synthesis).
5. The synthesis of alkylphosphonic acid esters by reaction of alkyl halides with trialkyl phosphites ($P(OR)_3$) (known today as the Michaelis-Arbuzov synthesis).

6. The first synthesis of alkylphosphonous dichlorides ($RPCl_2$) by the partial alkylation of phosphorus trichloride with an organometallic compound, and the similar synthesis of arylphosphonous dihalides.
7. The discovery of phosphonous acids (RPH(O)(OH)), where P is in a lower oxidation state.
8. Recognition of the rearrangement of tricovalent phosphorus compounds bearing an OH group to phosphoryl derivatives.
9. The synthesis of many new phosphines by reaction of his chlorophosphorus compounds with organometallic reagents.
10. The first synthesis of amides of C-P containing acids (phosphonic and phosphonous). (Michaelis did extensive work on other types of amides not containing a C-P bond; this work is not included in this Chapter).

3.3. THE SYNTHESIS AND REACTIONS OF ARYL PHOSPHONOUS DIHALIDES ($ArPX_2$)

The very first papers published by Michaelis on compounds with a C-P bond proved to be among his most significant [2]. Following the pioneering work of A. W. Hofmann during the period 1855 to 1873, the synthesis of any compound where phosphorus was attached to an aromatic ring remained to be accomplished (Hofmann had attempted a reaction of chlorobenzene with phosphonium iodide, PH_4I, a procedure he had developed for synthesizing alkyl phosphines, but it was unsuccessful [4]). Michaelis found that passing a gaseous mixture of benzene and phosphorus trichloride through a red hot tube, followed by condensing the gaseous mixture, gave a liquid product with much unreacted PCl_3 but on fractional distillation also a high boiling component that he called "phosphenyl chloride" throughout his career but which we call phenylphosphonous dichloride (also phenyldichlorophosphine). This compound boiled at 220°, and gave the correct C, H, and Cl analysis for $C_6H_5PCl_2$ (Eq. 1).

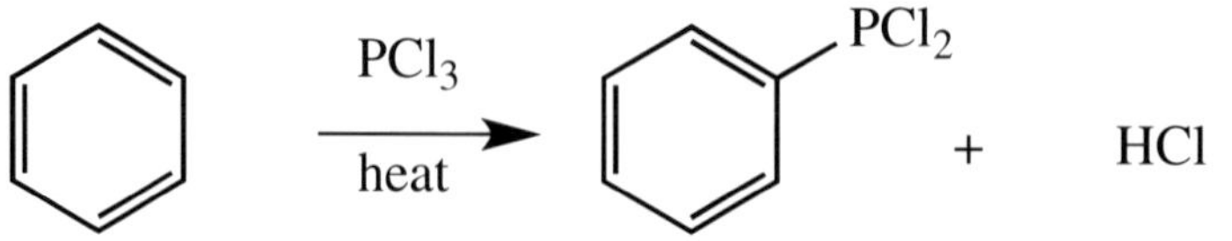

(1)

This process has been used as a practical industrial synthesis of phenylphosphonous dichloride, although undergoing some improvements. Thus, instead of a "red hot tube", a reactor temperature of 600-700° has been specified [5], and passing the gaseous mixture over catalysts [5b] (e. g., aluminum and $AlCl_3$) can reduce the reaction temperature to 100-150°. Michaelis published a lengthy review of the early literature of phosphorus chemistry (mainly that of A. W. Hofmann), and of his own work, which included further information on his benzene-PCl_3 reaction and provided diagrams of the apparatus he developed for the process [6]. He later tried to use toluene instead of benzene in the high-temperature process, but the results were poor and not useful [7]. He then proceeded to invent a different synthesis of aryl phosphonous dichlorides, again a process still in use today. This is a liquid-phase reaction of benzene and PCl_3 catalyzed by aluminum chloride, and thus resembling a Friedel-Crafts reaction [8]. Michaelis first showed that the product was identical to that from his gas-phase synthesis, and then applied it with success to toluene. The product was no doubt a mixture of isomers, dominated of course by the *para* isomer, but this mixture was not separated or analyzed and was reported to have a boiling point of 240°. (Kosolapoff [9] gives b. p. 245° for *para*, 235° for *meta*, and 244° for *ortho*). On further developing the reaction with toluene, Michaelis recognized that much of the phosphonous dichloride remained complexed to $AlCl_3$ and was not extractable or distillable. He did observe that adding small amounts of water broke up some of the complex and increased the yield [10]. Over the years the process has been improved by others and employed in the synthesis of aryl phosphonous dichlorides. It is of such value that the process for phenylphosphonous dichloride (with 72-78% yield) is included in *Organic Syntheses* [11]. In addition to the use of water, complex breakup by $POCl_3$ can be employed [12]. Michaelis showed in subsequent papers [13] that the process could be applied to a variety of aromatic hydrocarbons, including xylene, mesitylene, pseudocumene, cumene, and p-cymene. The process was also applied to benzene derivatives bearing halo and alkoxy groups. He also used dimethylaniline in this process [14], but it was later found that this amine so easily reacted with PCl_3 that no aluminum chloride was necessary to effect the reaction [15].

Michaelis discovered some fundamental properties of organophosphorus compounds on working with his "phosphenyl chloride" [2b]. He found that this compound readily added chlorine or bromine to form $PhPCl_4$ or $PhPCl_2Br_2$, respectively, and that the latter could add a second mole of bromine to form $PhPCl_2Br_4$. With oxygen, $PhPCl_2$ was converted to the oxychloride, $PhPOCl_2$. This type of compound had been prepared first by Hofmann from the reaction of alkylphosphonic acids with PCl_5. Phenylphosphonous dibromide was first prepared by Michaelis on reacting $PhPCl_2$ with dry HBr [16]. However, he also devised another route to $PhPBr_2$, the monophenylation of PBr_3 with diphenylmercury, a reaction type he later used in the synthesis of alkylphosphonous dichlorides (see Section 3.5) and of tertiary phosphines (Section 3.6).

During these studies Michaelis became the first to synthesize a member of the family of phosphonous acids, of molecular formula RPO_2H_2 [17]. He accomplished this by careful hydrolysis of $PhPCl_2$, and named the compound "phosphenyligesäure", known today as phenylphosphonous acid (Eq. 2). He recognized that it was monobasic (**2**) and not the expected dibasic form (**1**) (this important property is discussed further in Section 3.9).

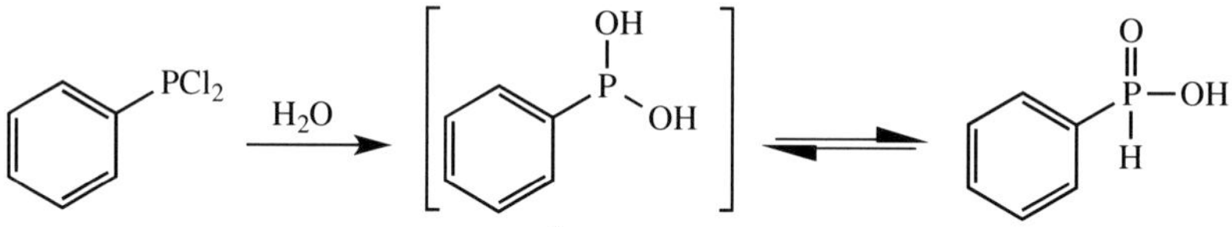

1 2 (2)

On working with this compound, he encountered the important parent phosphine $PhPH_2$ for the first time, recognizing it as the phosphorus counterpart of aniline. He had earlier attempted to make this compound by the reduction of phenylphosphonous dichloride with zinc and HCl but with little success. Nowadays we can accomplish this reduction with ease using metallic hydrides, but these were not in use in Michaelis's day. Phenylphosphine (**3**) was formed on heating the acid (**2**) in what is now described as a disproportionation reaction [18], which also gives the higher oxidation state compound phenylphosphonic acid (**4**), (Eq. 3). Phenylphosphine is quite unlike aniline; it is very difficult to work with because of its notoriously foul smell and ready oxidation on exposure to air. It is much weaker as a base, failing to form salts with aqueous acids. Michaelis did, however, prepare a salt with anhydrous HI.

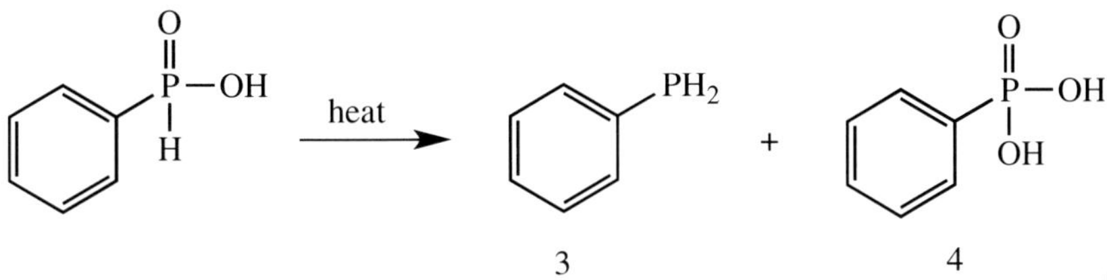

(3)

Michaelis found that in his "phosphenyligersäure" (**2**) there must be a difference in the two hydrogens, since it only had the properties of a monobasic acid. He showed this by making a number of metallic salts, whose analysis proved that the anion had only one negative charge [17]. He proposed that one hydrogen was attached to phosphorus, the other (the acidic one) to oxygen, and likened the situation to that in phosphorous acid (H_3PO_3), which is dibasic, not the expected tribasic [19]. He drew his phosphonous acid as $OPH(OH)C_6H_5$ and phosphorous acid as $OPH(OH)_2$, both acceptable today, although failing to express the multiple bonding of O to P [17]. These observations, confirmed over the years with many other examples, led Michaelis to predict a very important generality in phosphorus chemistry, that tricovalent phosphorus bearing an OH group isomerizes by hydrogen migration from O to P. This is now known to be an equilibrium process, heavily dominated by the PH form. It thus resembles the keto-enol tautomeric equilibrium of carbonyl compounds. Only in special cases is this generality invalid [20a].

Numerous other valuable reactions with phenylphosphonous dichloride were discovered, resulting in the synthesis of some structures novel at the time. Most have become standard methods of synthesis over the years. These are summarized below.

3.A. With Alcohols

It was found that towards ethanol $PhPCl_2$ acted primarily as a chlorinating agent (Eq. 4) [21].

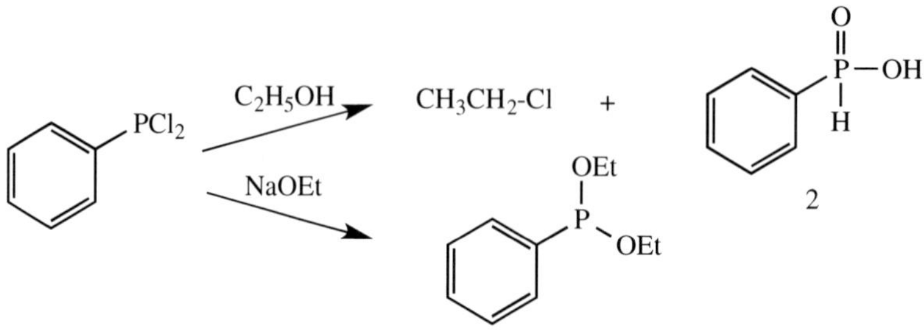

5 (4)

However, the reaction with sodium ethoxide in ether gave a new type of compound, $PhP(OEt)_2$ (**5**) which Michaelis called a “phosphenyläther”, now known as diethyl phenylphosphonite. Hydrolysis gave phenylphosphonous acid. Similarly, amines displaced chlorine to form amides of general structure $PhP(NR_2)$ [22].

3.B. With Sulfur

On heating with sulfur, phenylphosphonous dichloride was converted to the thiophosphoryl derivative $PhP(S)Cl_2$. The sulfur was retained on hydrolysis with aq. KOH, forming the potassium salt of formula $PhPSO_2K_2$. With ethanol, the ester $PhP(S)(OEt)_2$ was formed [23].

3.C. With Organometallics to Synthesize Tertiary Phosphines

In 1857 Cahours and Hofmann had shown that alkylzinc compounds could alkylate phosphorus trichloride and thus effect the synthesis of trialkylphosphines [28]. Michaelis used this type of reaction most successfully for the synthesis of tertiary phosphines in the aromatic series by reacting his arylphosphonous dichlorides first with dialkyzinc compounds (Eq. 5), and later with the less reactive dialkylmercury compounds. The first phosphines he reported were diethylphenylphosphine and dimethylphenylphosphine [6,25]. He noted that the phosphines formed a complex with the zinc chloride generated in the reaction, but that this could be broken with aqueous KOH. Other phosphines, and some new reactions performed on phosphines, will be described in Section 3.6.

Michaelis also used his phosphonous dichlorides as precursors of triarylphosphines. His was the first practical synthesis of triphenylphosphine, which today is probably the best known of organophosphorus compounds. His method consists of the reaction of phenylphosphonous dichloride with bromobenzene and sodium (probably acting as C_6H_5Na) in ether, with a reaction time of 6-10 days at room temperature [26]. He later obtained triphenylphosphine by this method as applied to phosphorus trichloride [27]. He went on to use this method for the synthesis of several triarylphosphines, including tri-p-tolyl, tri-m-xylyl, tri-p-xylyl, tripseudocumyl, and trimesityl, described in Section 3. 6.

3.4. Synthesis and Reactions of Diarylphosphinous Chlorides

Michaelis showed in 1875 that his procedure of reacting phosphorus trichloride with diphenylmercury [28] could be applied to phenylphosphonous dichloride to introduce a second phenyl on phosphorus. The product would have been diphenylphosphinous chloride, but this was not isolated at this time; the reaction product was directly subjected to hydrolysis/oxidation with nitric acid to form diphenylphosphinic acid (Eq. 6) [28].

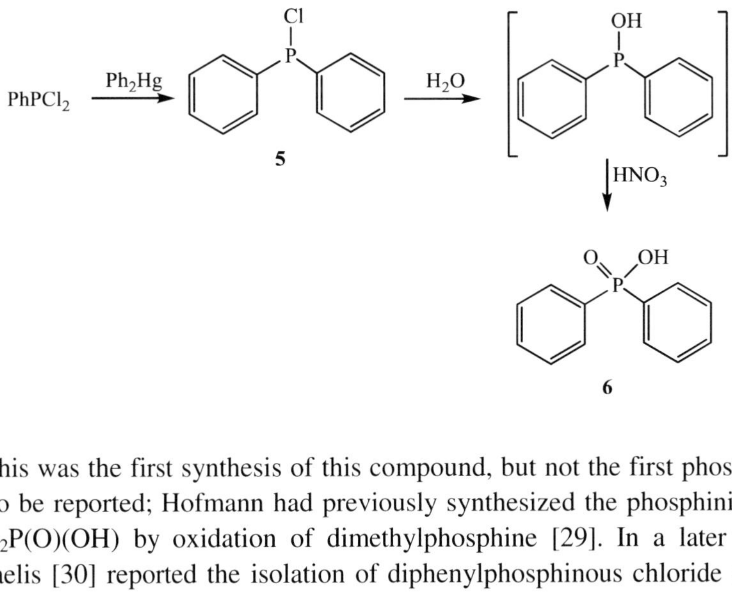

(6)

This was the first synthesis of this compound, but not the first phosphinic acid to be reported; Hofmann had previously synthesized the phosphinic acid $(CH_3)_2P(O)(OH)$ by oxidation of dimethylphosphine [29]. In a later paper Michaelis [30] reported the isolation of diphenylphosphinous chloride (**5**) by distillation of the $PhPCl_2$- Ph_2Hg reaction product. In an accompanying paper from the Karlsruhe laboratory [31], it was reported that the phosphinous chloride could be obtained by thermal disproportionation of phenylphosphonous dichloride at 280°, a still-useful process. (Eq. 7).

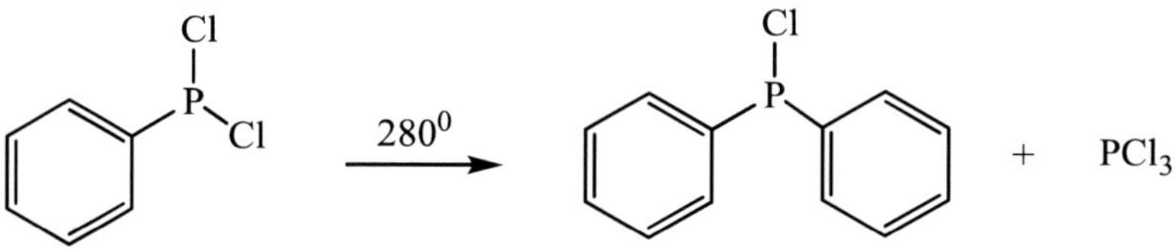

5

(7)

Michaelis confirmed that with nitric acid hydrolysis of Ph_2PCl occurred and the intermediate (Ph_2PHO, not isolated; see below) was oxidized to yield the same monobasic compound $Ph_2P(O)(OH)$ (**6**) observed earlier [30]. He also found that the chloride reacted with chlorine to give the trichloride Ph_2PCl_3, which could be hydrolyzed to diphenylphosphinic acid. In a later paper it was shown that the acid **6** could be converted to the phosphinic chloride ($Ph_2P(O)Cl$) with PCl_5 (Scheme 1) [32].

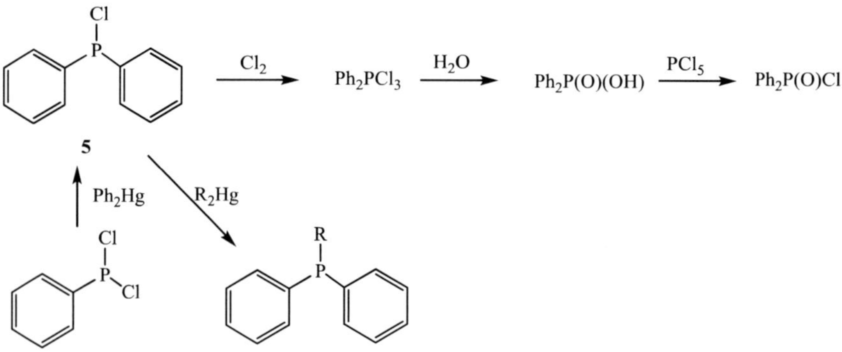

Scheme 1.

Michaelis also described the synthesis of Ph_2PCl by the reaction of $PhPCl_2$ with diphenylmercury [33], and in the same paper he reported the synthesis of tertiary phosphines by the alkylation of Ph_2PCl. (Scheme 1).

Several other new compounds were derived from diphenylphosphinous chloride. The hydrolysis product, written as Ph_2POH but later proved to have the structure $Ph_2P(O)H$ from the same rearrangement as discovered for phosphonous acids (see Section 7.D), was shown to be a source of diphenylphosphine (**7**) from a disproportionation reaction (Eq. 8) that occurred from the heat released in the reaction, but also if the hydrolysis was conducted in NaOH solution [26]. This new structure was only weakly basic and was not soluble in dilute acids.

$$\underset{}{Ph_2P(O)H} \xrightarrow{\text{heat}} \underset{\mathbf{7}}{Ph_2P\text{-}H} + \underset{\mathbf{6}}{Ph_2PO(OH)} \qquad (8)$$

The chlorine was also displaced by phenoxide ion to produce the first known phosphinite ester (Eq. 9) [32]. This compound also on hydrolysis gave "Ph_2POH" which proceeded to undergo the disproportionation reaction. The phosphinite was oxidized by air to form $Ph_2P(O)(OPh)$, and with sulfur or selenium to form $Ph_2P(S)(OPh)$ and $Ph_2P(Se)(OPh)$, respectively, all new ester types.

$$Ph_2P\text{-}Cl \xrightarrow{NaOPh} Ph_2P\text{-}OPh \xrightarrow{O_2} Ph_2P(=O)\text{-}OPh \qquad (9)$$

The phosphorus atom was alkylated with methyl iodide or benzyl chloride to form stable salts which on hydrolysis gave mixed tertiary phosphine oxides (Eq. 10). Diphenyphosphinous chloride was also noted to undergo alkylation with benzyl chloride to form $Ph_2P(OR)(CH_2Ph)^+ Cl^-$, which on hydrolysis gave the phosphine oxide (Eq.10). Trialkyl phosphine oxides had been previously observed by Hofmann [24]. This is an unusual alkylation reaction in that other types of trivalent phosphorus-chlorine compounds ($ArPCl_2$, $RPCl_2$, R_2PCl) are not known to form adducts with alkyl halides.

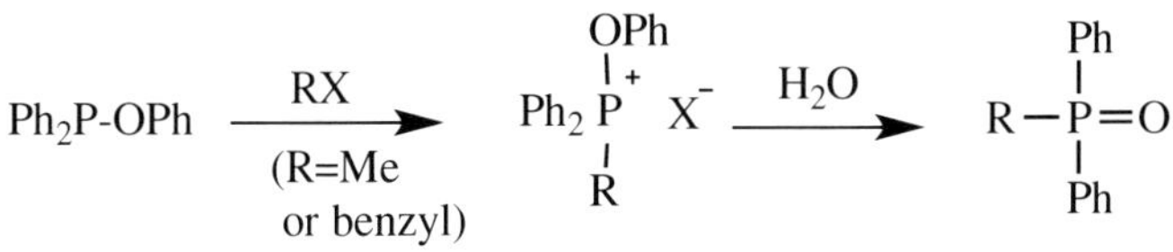

(10)

3.5. The Synthesis and Reactions of Alkylphosphonous Dichlorides

Although Hofmann had earlier done pioneering work on primary, secondary and tertiary alkylphosphines, the important family of alkyl chlorophosphines remained unknown until Michaelis developed a method for their preparation. He accomplished this by using dialkyl mercury derivatives to alkylate phosphorus trichloride [34]. An obvious difficulty in the alkylation of PCl_3 is that the monoalkylation product $RPCl_2$ is itself subject to alkylation, which occurs readily with the more reactive organometallics. In fact, Hofmann had previously [24] used dialkylzinc compounds in reaction with PCl_3 to obtain tertiary phosphines. Organomercury compounds are among the less reactive alkyl transfer agents, which made possible the monoalkylation of PCl_3 (Eq. 11).

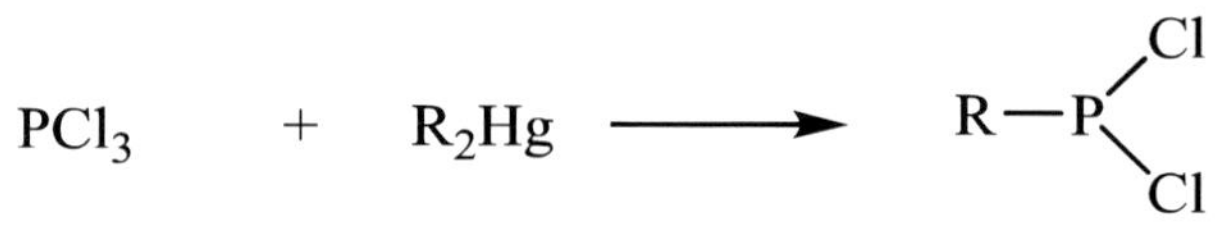

(11)

The reactions were performed in sealed tubes at 100 ° or higher. He synthesized ethyl, isopropyl, and isoamyl phosphonous dichlorides by this method. He proved that he had prepared ethylphosphonous dichloride by hydrolyzing and oxidizing the compound in the air to form ethylphosphonic acid, with properties like those described by Hofmann for this compound when made from the oxidation of ethylphosphine. Michaelis knew that the lower oxidation state phosphonous acids (known to him as "phosphorigesäurer") must be precursors of the phosphonic acids, but did not attempt to isolate these intermediates (Scheme 2). Ethylphosphonous acid (and other acids) was not described until 1890 in a paper from Rostock by F. Guichard [35], not bearing the name Michaelis but probably associated with his program. Guichard provided much new information on the alkylphosphonous dichlorides made by the Michaelis method, and on their properties.

Curiously, no mention is made in Michaelis's papers, nor in that of Guichard, about the parent of the series, methylphosphonous dichloride; this important compound remained unknown until its preparation was mentioned as a brief note in 1958 [36c]. For a large-scale synthesis, see reference [36b]. It is now a well-known chemical

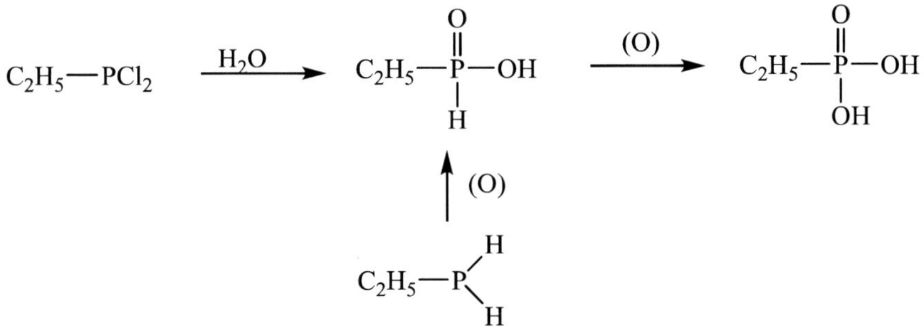

Scheme 2.

Michaelis prepared the first alkyltetrachlorophosphorane by adding chlorine to $EtPCl_2$ to form $EtPCl_4$ and on partial hydrolysis this compound provided ethylphosphonic dichloride. Hofmann had been the first, in 1871, to synthesize a member of this family, CH_3POCl_2, by the reaction of methylphosphonic acid with PCl_5 [29].

Guichard performed a reaction of isoamylphosphonic acid with isoamylphosphonic dichloride and isolated a crystalline product, m.p. 122°, with the empirical formula $C_5H_{11}PO_2$ (Eq. 12).

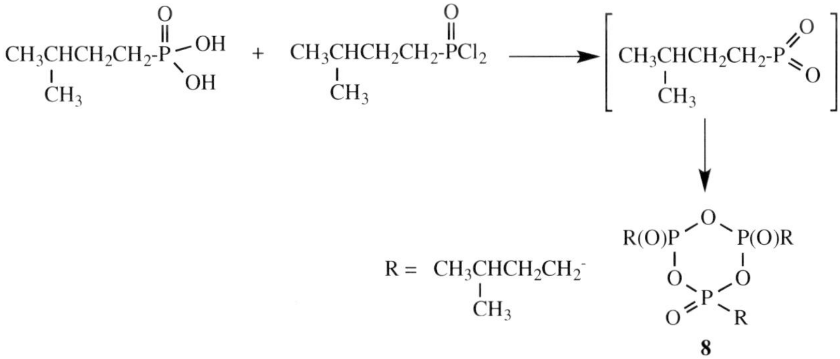

8

(12)

He claimed this formula represented the anhydride of isoamylphosphonic acid. In modern terminology, that structure would fall into the class of dioxophosphoranes, where P has three coordinated (bonded) atoms (designated σ^3; for a discussion of this notation, see reference 20b) and a total of five bonds (designated λ^5). However, no simple P-alkyl or P-aryl dioxophosphorane structure has ever been isolated [37a]. Instead, cyclic trimers or polymers are formed. Guichard's compound is indeed an anhydride but most likely a cyclic trimeric one, not monomeric, with structure **8**. Further, as noted also in Section 6, simple phosphorus compounds in general do not exist with any type of *pi* multiple bond; such structures were unknown until about the 1970's when it was discovered that they could be stabilized by very large substituents, or in some cases by strongly electron-withdrawing groups. Similarly, Michaelis assumed that amide derivatives (e. g., Et_2N-PO_2 [38]) of the unknown anhydride metaphosphoric acid (HO-PO_2) had the stable monomeric structures, and that anhydrides of arylphosphonic acids had the structure $ArPO_2$. [13, 39-42]. However these too are known to have but fleeting existence and in general structures with $\sigma^3\lambda^5$ bonding form cyclic or polymeric structures. As an elaboration on unsaturated phosphorus derivatives, compounds with phosphorus in the lower oxidation state ($\sigma^2\lambda^3$) would require a true *pi* bond; they are generally not stable and form dimeric or polymeric structures [41b]. Michaelis recognized that the compound PhN=PCl that he prepared from aniline and PCl_3 had a higher molecular weight than expected (found, 375; calculated, 325) and he drew a dimeric structure as a possible contributor to account for this [43a]. The compound PhNH-N=PPh that he prepared from phenylhydrazine and $PhPCl_2$ is probably also dimeric [43b].

3.6. The Synthesis of Phosphines from Chlorophosphorus Compounds; Derivatives Prepared from the Phosphines

It was known when Michaelis started his research, due to the work of Cahours and Hofmann [24], that phosphorus resembled nitrogen in forming structures with three alkyl groups of attachment. When his aryl phosphorus chlorides, and later alkyl phosphonous dichlorides, became available, he applied a reaction known to replace chlorine of phosphorus trichloride with carbon substituents, namely, the reaction with organometallic reagents. Most of his work made use of organomercury compounds, but he also used the more reactive zinc compounds. He was thus able to prepare, from aryl phosphonous dichlorides, the first phosphines with mixed P-aryl and P-alkyl groups. The compounds prepared were:

$PhPMe_2$ [6,25]
$PhPEt_2$ [6,25]
p-Me_2N-C_6H_4-PPh_2 [14a]

p-Me_2N-C_6H_4-PMe_2 [14b]
Ph(p-MeC_6H_4)PEt [40]

He also prepared triaryl phosphines by arylation of PCl_3 with aryl halides and sodium (thus ArNa). A list is provided in Ref. 40 and includes:

Tri-p-tolylphosphine
Tri-m-xylylphosphine
Tri-p-xylylphosphine
Tripseudocumylphosphine
Trimesitylphosphine

Triphenylphosphine deserves special mention because of the importance of this compound. It was first reported [26] by the phenylation procedure applied to $PhPCl_2$, but as noted above an improved procedure consisted of the phenylation of PCl_3 [27a]. Phosphines with mixed aryl substituents, including diphenyltolyl, phenyldi-tolyl, and chlorophenylditolyl, were also reported in this paper:

From Ph_2PCl were prepared the tertiary phosphines Ph_2PCH_3 and $Ph_2PC_2H_5$ by alkylation with zinc alkyls [33], and from these phosphines he prepared quaternary phosphonium salts by the reaction with alkyl halides, a process dating back to the earliest days of phosphorus chemistry (as discussed in Ref. 1).

Hofmann was the first to show that tertiary phosphines could be oxidized to give the new family of phosphine oxides, in particular preparing triethylphosphine oxide, $Et_3P{=}O$ [24]. Michaelis applied this reaction to his aryl-containing tertiary phosphines and synthesized several new phosphine oxides. He performed the first electrophilic substitution of these compounds; he reacted triphenylphosphine oxide with a nitric acid-sulfuric acid mixture and prepared tri(m-nitrophenyl)phosphine oxide as a yellow solid [27b, 44]. He did not show the orientation of the nitro groups in these papers, but years later the meta-orientation was confirmed by others, and is shown in reference [9]. He went on to reduce the nitro groups with tin and HCl to give the first phosphorus compound containing the amino group, tri(m-amino-phenyl)phosphine oxide. He proceeded to alkylate the amino groups with methyl iodide in alcohol at 100° (Eq.13). He also acetylated and benzoylated his tris-aminophenyl compound [27, 44]. In addition, he prepared a trinitro derivative from tri-p-tolylphosphine oxide in this work.

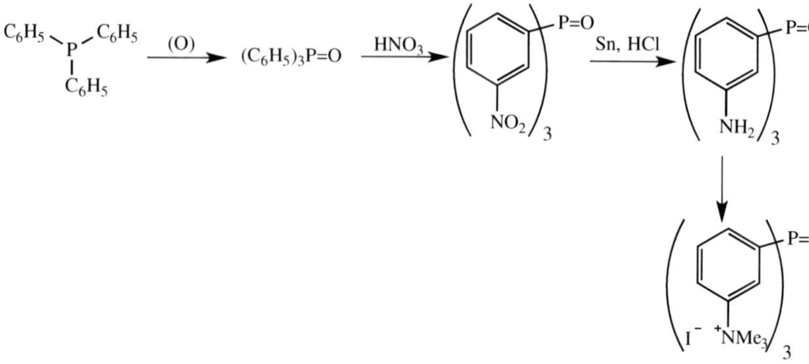

(13)

In another approach to functionalized tertiary phosphines, Michaelis oxidized the three methyl groups of tri(p-tolyl)phosphine oxide with chromic acid to obtain the structure $P(O)(p\text{-}C_6H_4\text{-}COOH)_3$ [40].

Michaelis devised a new method for converting phosphines to phosphine oxides. He found that tertiary aromatic phosphines added chlorine smoothly to form a new species Ar_3PCl_2, and that on hydrolysis this gave the same phosphine oxide as direct oxidation of the phosphine [27b].

Hofmann had found that his tertiary alkylphosphines formed salts with acids, coordination complexes with $PtCl_2$, and added elemental sulfur to form phosphine sulfides [24]. He prepared the first quaternary phosphonium salts by alkylating his phosphines with alkyl iodides, and from these salts he prepared with moist Ag_2O (acting as "AgOH") the first quaternary phosphonium hydroxides. All of these reactions were applied by Michaelis to his tertiary aryl-substituted phosphine, as reported in various references cited above.

August Hofmann had included ethyl chloroacetate among the alkyl halides he had reacted with triethylphosphine, and obtained the expected quaternary salt. He then removed chloride and the ethoxy group of the ester by treatment with moist Ag_2O. He correctly referred to the product as a betaine, a term in use for the corresponding nitrogen compound. Structural theory at that early time did not lead him to propose a dipolar structure for the product and he presented it simply in molecular form [45a], as did Meyer for the similar product prepared from trimethylphosphine [45b]. Michaelis also used ethyl chloroacetate in his alkylation studies of aryl phosphines [40.46]. He referred to these and other alkyl halide reaction products as salts, but showed structures with halogen directly bonded to P. The salt **9** from tri-p-tolylphosphine was hydrolyzed to give the betaine, the term Michaelis used but he represented it with cyclic structure **11** rather than the correct dipolar form (**12**, Scheme 3). All of his 5-coordinate phosphorus salt structures we now know to need clarification; 5-coordinate phosphorus is only stable in special cases, and there is no evidence that this is one of them. In fact the first authentic 5-coordinate phosphorus compound (Ph_5P) was not obtained until 1947 [47].

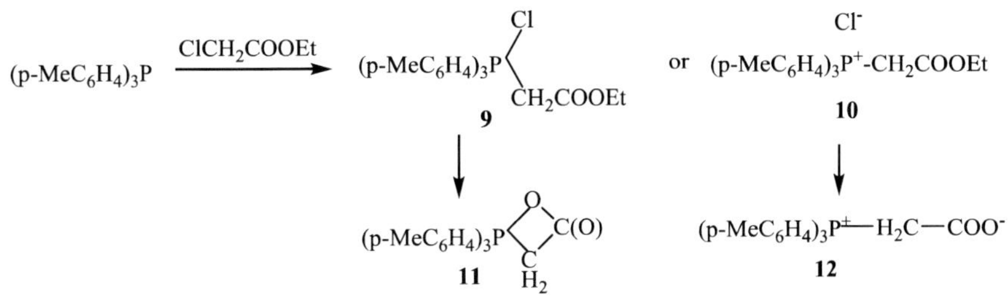

Scheme 3.

Similarly, Michaelis proposed the structure **13** for the reaction product of tri-p-tolylphosphine with chloroacetone [40], and **14** for the product of its hydrolysis with sodium carbonate (Eq. 14). Curiously, Michaelis referred to **15** as "tritolylphosphorketobetain", but it cannot have a betaine structure, and one wonders what structural features Michaelis actually meant here for the term betaine (Eq. 14).

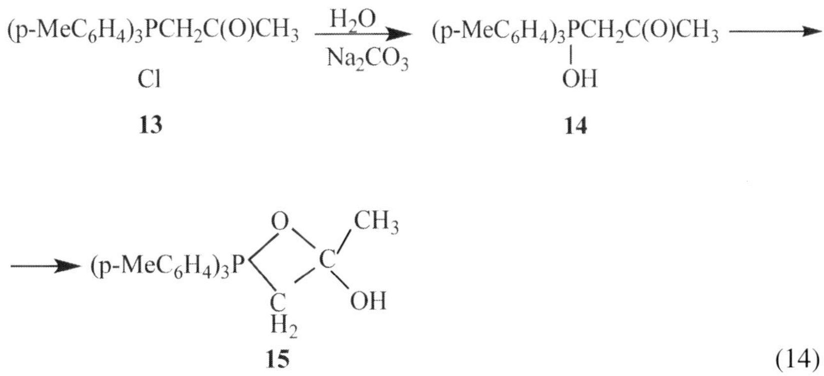

(14)

Another early structural mis-assignment of Michaelis might be put forward, again one that is understandable in his day. He proposed structures with P to P double bonds, but the first authentic compounds of this structure were not prepared until 1981 [48]. It was noted in Section 3.5 that only in very special cases does phosphorus form *pi* double bonds. Thus for the product derived from phenylphosphonous dichloride with phosphine he proposed the structure Ph-P=P-OH [25,49], and PhP=PPh for the product from $PhPCl_2$ and $PhPH_2$. The latter compound was re-synthesized by Kuchen and Buchwald [50] and found to be a cyclic dimer; a cyclic structure is also probable for Michaelis's proposed compound PhP=P-OH.

3.7. Synthesis and Properties of P-Aryl Phosphorus Acids

Michaelis's success in synthesizing phosphonous dichlorides and phosphinous chlorides bearing aryl substituents was the starting point for a great amount of work on the various types of phosphorus acids. The attempt will be made here to organize the results given in his many papers in terms of the type of acid functional group studied.

7.A. Phosphonic Acids

Michaelis referred to the parent $PhPO(OH)_2$ first as "phosphenylsäure", later using the term "phosphinic" acids for the family, thus creating the name "phenylphosphinic acid". In modern usage, the proper name is phenylphosphonic acid. He recognized the similarity to the alkylphosphonic acids first reported by Hofmann in 1872 [51]. His first synthesis with a characterization of $PhPO(OH)_2$ appeared in 1873 [2b]; it had been mentioned with no details in reference [2a] and further studied in reference [2c]. He obtained this compound from reactions of $PhPCl_4$, prepared as noted in Section 3.3 by the addition of chlorine to $PhPCl_2$, with water or oxalic acid and reported a melting point of 158° (later confirmed by several others). The acid was converted to the silver salt, giving the proper silver analysis for $PhPO(OAg)_2$ which proved the dibasic character of the acid [52]. He also found that oxygen reacted with $PhPCl_2$ to form the oxychloride, $PhPOCl_2$, which gave phenylphosphonic acid on hydrolysis. Yet another approach to this acid was reported later [6], consisting of the oxidation of "phenyl phosphenyligesäure" (PhPH(O)(OH), our phenylphosphonous acid), obtained by hydrolysis of $PhPCl_2$. In Scheme 4 these various reactions are summarized; together they express some fundamental relationships of organophosphorus chemistry that we still depend on today.

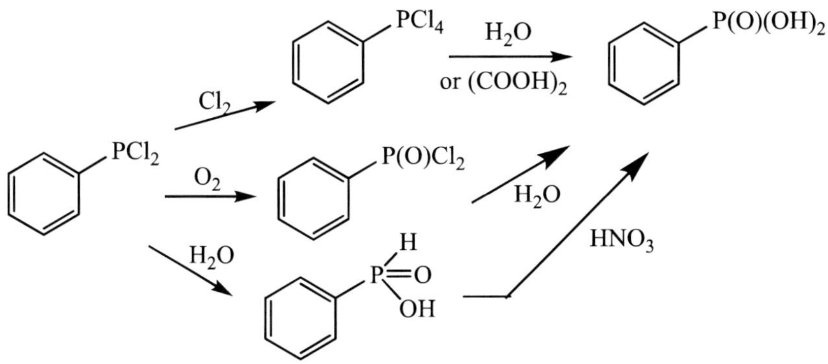

Scheme 4.

Phosphonic acids were found to be quite stable, but Michaelis did note that under the strenuous conditions of heating at 250° the C-P bond was broken and benzene and "metaphosphoric acid" (said to be PO_3H, but in fact a polymer as discussed in Section 3.5) were formed. At the lower temperature of about 200°, water was lost and polymeric phosphoric anhydrides were formed.

Using some of the above reactions, Michaelis went on to prepare several other arylphosphonic acids. He had developed the aluminum chloride catalyzed reaction of aromatic hydrocarbons with PCl_3 by this time, gaining access to new phosphonous dichlorides for use as starting materials. The following phosphonic acids were prepared: p-tolyl [8, 10], o-tolyl [52], m-tolyl [52], p-ethylphenyl [52], mesityl [13], 2,4,5-trimethylphenyl (pseudocumyl) [13]. Phosphonic acids were also prepared starting with cumene and cymene [13], as well as with diphenylmethane [40], but the position of the acid group in these was uncertain. Some of these structures are illustrated in Scheme 5. Aromatic compounds with certain substituents on the benzene ring other than alkyl groups also gave phosphonous dichlorides with PCl_3 and $AlCl_3$ and these were converted to arylphosphonic acids. Prepared were p-dimethylamino [14a], p-chloro [39], p-bromo [39] and p-methoxy [39].

With synthetic methods for arylphosphonic acids in hand, Michaelis proceeded to study their properties, applying conventional methods of substitution and functional group modification, and many new acids resulted from this work. Some of these are summarized in Schemes 6 and 7.

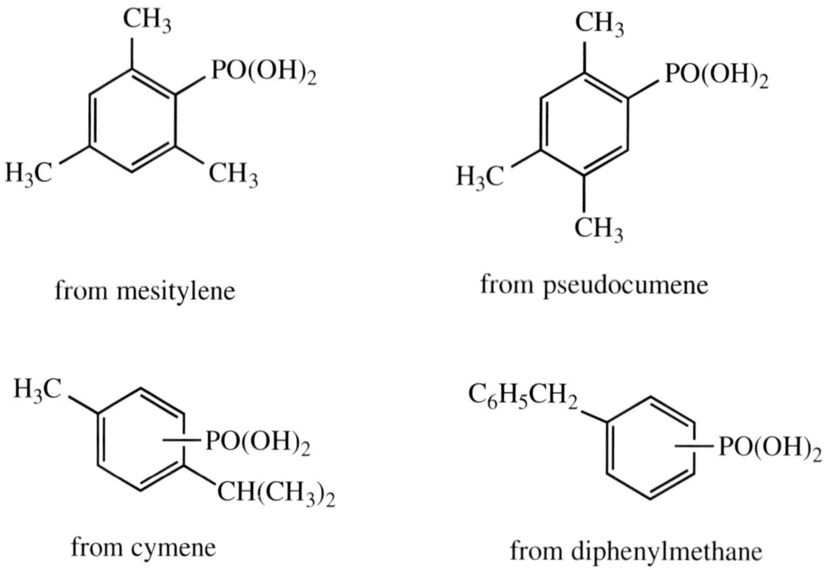

Scheme 5.

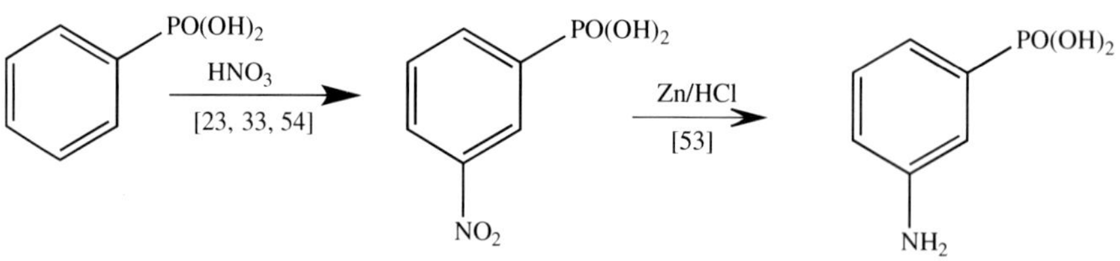

Scheme 6.

A number of new compounds were prepared by replacement of Cl in p-tolyl-$POCl_2$ by amino or phenoxy groups to give amides or esters, respectively [52] (Another paper describing reactions of $PhPOCl_2$ with amines [52c] contains some structures with P=N bonds, which are incorrect; probably the structures are cyclic dimers). For conversion of the phosphonic acid group to phosphonates, Michaelis made silver salts and then reacted these with alkyl iodides. He observed that Fischer acid-catalyzed reaction of the phosphonic acids with alcohols failed, an observation of a generality that became universal in phosphorus acid chemistry. He prepared phosphine oxides by reaction of the $POCl_2$ group with dialkyl mercury compounds. (Scheme 8.)

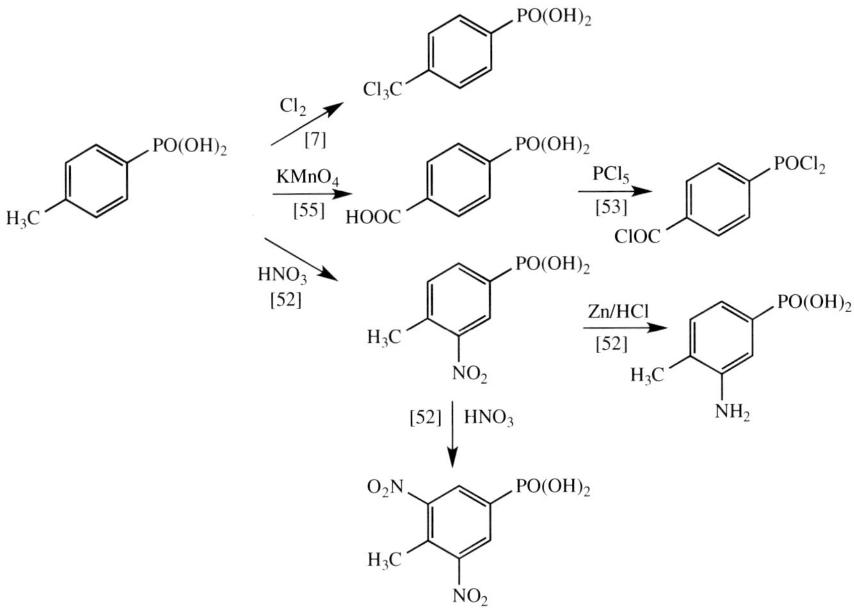

Scheme 7.

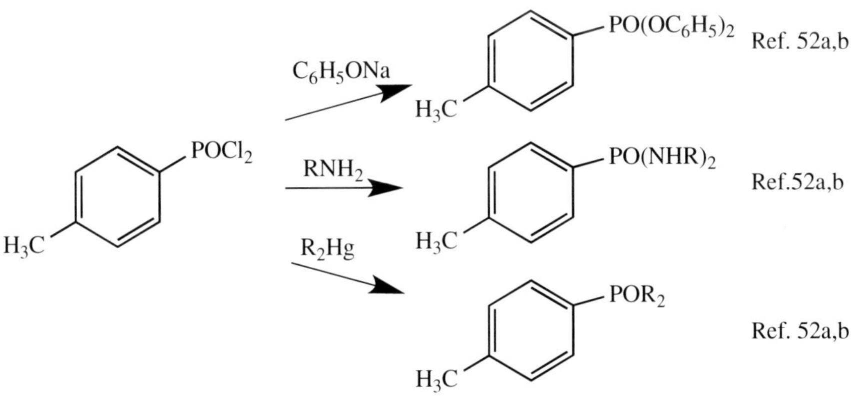

Scheme 8.

Michaelis also prepared phosphonic acids with ortho-tolyl and meta-tolyl substituents, using his reaction of diaryl mercury compounds with phosphorus trichloride to obtain the tolylphosphonous dichlorides needed as starting materials, and then conversion to the phosphonic acids by the methods outlined in Scheme 4 [52]. From these he prepared, among others, the nitro derivatives, the amino derivatives, the chloro derivatives, and the carboxy derivatives. In addition he prepared derivatives of p-ethylphenylphosphonic acid and of 2,5-dimethylphosphonic acid (starting with the phosphonous dichloride derived from p-xylene) [52].

7.B. Phosphinic Acids ($Ar_2PO(OH)$)

Related to forming phosphonic acids from $ArPCl_2$ as starting materials, Michaelis employed the phosphinous chlorides, Ar_2PCl, for the synthesis of diarylphosphinic acids. The same reactions shown in Scheme 4 were used to prepare both types of acids, namely, the addition of chlorine and hydrolysis of Ar_2PCl_3, or oxidation to prepare $Ar_2P(O)Cl$ then hydrolysis, or oxidation of $Ar_2P(O)H$.

Diphenylphosphinic acid was the first to be prepared [28,30], using the reaction of $PhPCl_2$ with Ph_2Hg to obtain the phosphinous chloride. This was hydrolyzed to $Ph_2P(O)H$ and then oxidation performed. Such methods were also used to prepare di-p-tolyl- and di-pseudocumylphosphinic acids [13]. Other phosphinic acids with two different aryl groups were prepared by reacting phenylPCl_2 or tolylPCl_2 with diaryl mercury compounds, then converting the resulting phosphinous chlorides (ArAr'PCl) by the usual reactions to ArAr'PO(OH). Prepared were phenyltolylphosphinic acid, phenylpseudocumylphosphinic acid, and others [40].

Michaelis devised other methods to synthesize phosphinic acids. In one of these, he used $POCl_3$ as the starting material, a much more readily available compound than $ArPCl_2$. Reaction with one equivalent of piperidine ($C_5H_{10}NH$) as a blocking agent gave the compound $(C_5H_{10}N)POCl_2$, which then was reacted with two equivalents of an aryl Grignard reagent (a great improvement over the use of the less reactive mercury compounds) to give diarylphosphinic amides. Hydrolysis provided the phosphinic acids [56]. Several acids were prepared by this method, including di-p-tolyl, di-o-tolyl, dibenzyl, and dinaphthyl (this paper was published in 1915 during World War I; Michaelis made the poignant comment that he could probably make alkylarylphosphinic acids by this method but the war interfered with his work. This appears to be his last paper on phosphorus acids; he died the following year). Eqn 16:

$$POCl_3 \xrightarrow{(CH_2)_5NH} (CH_2)_5N\text{-}POCl_2 \xrightarrow{RMgX} (CH_2)_5N\text{-}P(O)R_2 \xrightarrow{H_2O} R_2P(O)(OH)$$

(R=o- and p-tolyl, benzyl, naphthyl)

An earlier method [22] had also made use of amide derivatives as starting materials, and by this method he did prepare alkylarylphosphinic acids. Here he used the di-piperidide from reaction of piperidine with $PhPCl_2$ and then alkylated phosphorus with an alkyl iodide. This gave the phosphonium derivative $RPh(pip)_2P^+ I^-$; iodide was replaced by OH^- by the Hofmann method [24] of reacting the salt with moist silver oxide. Hydrolysis then gave the phosphinic acid (Eq. 16). Prepared were methylphenyl- and ethylphenylphosphinic acids.

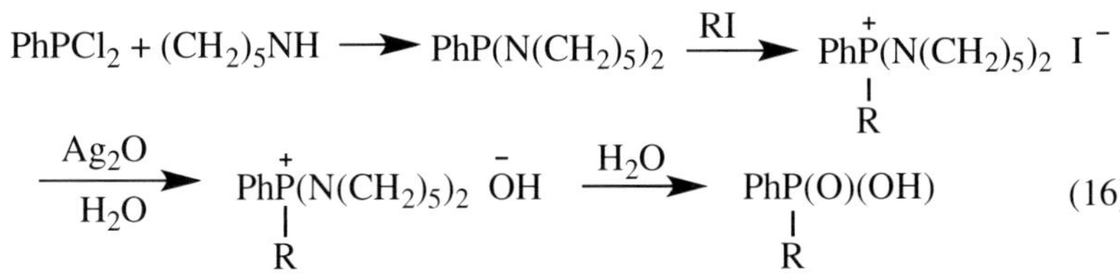

(16)

In a similar manner, diphenyl esters of phosphonous acids ($ArP(OPh)_2$), prepared from $ArPCl_2$ and sodium phenoxide [40], were alkylated with alkyl iodides to give stable crystalline phosphonium salts (**16**, Scheme 9). However, when this reaction was performed on alkyl esters of phosphonous acids (**17**), a different reaction path to phosphinates was followed: the intermediate phosphonium salts (**18**) were unstable and not directly observed. Their decomposition is the result of the halide ion attacking carbon of the alkoxy group in an S_N2 reaction (Scheme 9). As will be discussed in the next section, related behavior occurs in the famous Michaelis-Arbuzov synthesis of phosphorus acids. On hydrolysis with sodium hydroxide, the stable salts (**16**) from diphenyl esters give alkylarylphosphinates. Thus from the ester $PhP(OPh)_2$ with methyl iodide was prepared methylphenylphosphinic acid; p-tolylP$(OPh)_2$ with methyl iodide led to methyl-p-tolylphosphinic acid, and with benzyl chloride to benzyl-p-tolylphosphinic acid.

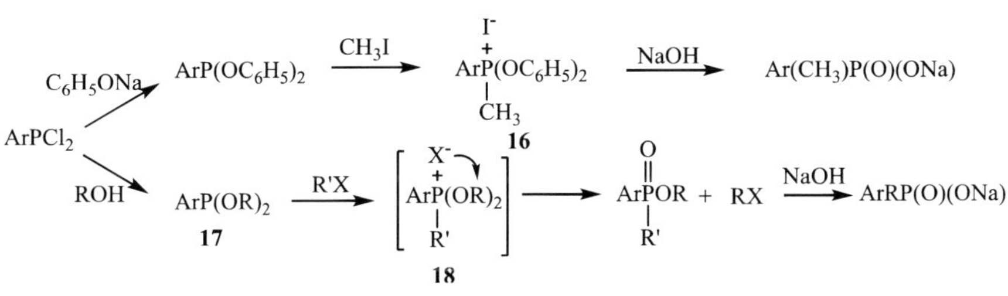

Scheme 9.

Other phosphinic acids were prepared from those mentioned above [40]. Thus, nitration of di-p-tolylphosphinic acid gave the di-nitro compound **19**. On permanganate oxidation, this acid was converted to the mono-carboxy derivative **20.** Phenyl-p-tolylphosphinic acid was nitrated in both rings to give the dinitro derivative **21**.

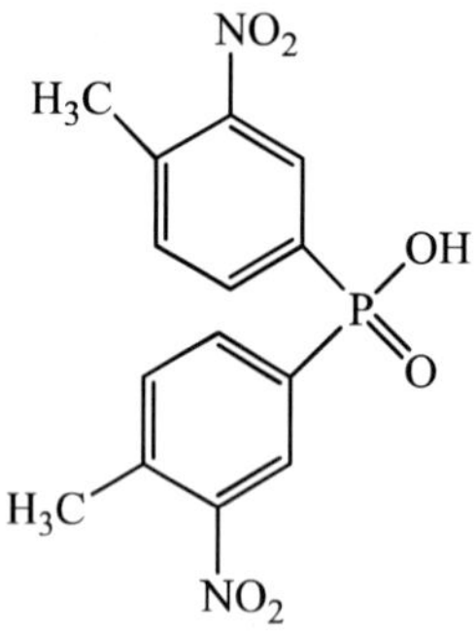

19

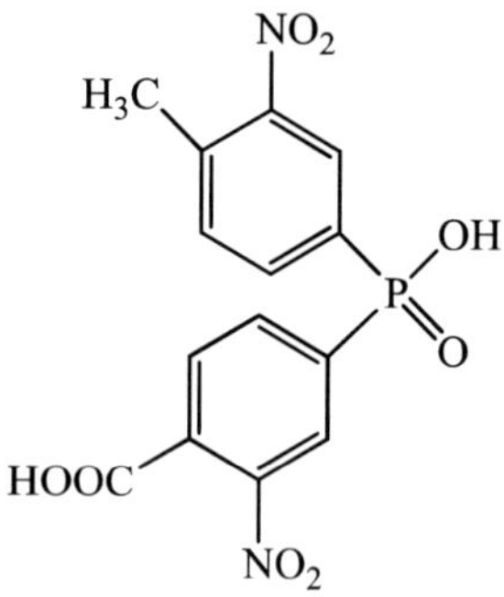

20

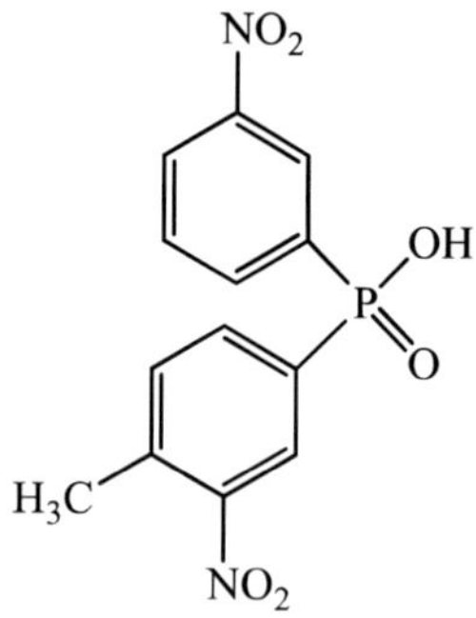

21

7.C. Phosphonous Acids (ArPH(O)(OH))

These acids were referred to first as "phosphenyliger" acids [49] and later as "phosphiniger" acids (e.g. [57]. The major method of synthesis in the aryl series is by acid hydrolysis of arylphosphonous dichlorides ($ArPCl_2$), as has been described in Section 3.3. Michaelis [6] noted that there was only one acidic OH group in these compounds and proposed that the proper structure had one hydrogen on phosphorus as shown previously by structure **2 a**nd was not the di-hydroxy form **1.** This point is discussed further in Section 3.9.c.

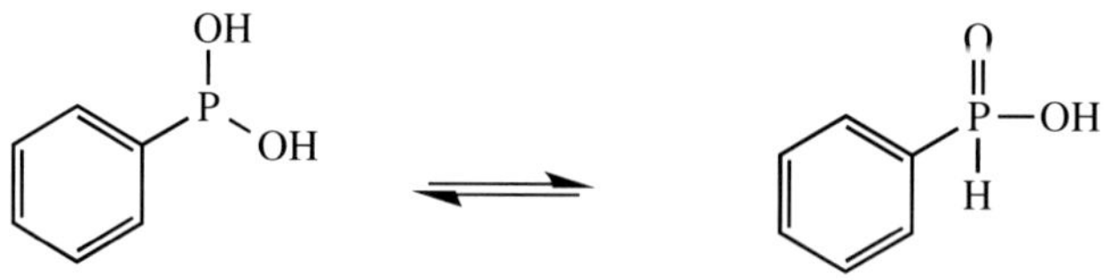

1 **2**

Michaelis prepared numerous P-aryl acids by this method; these are shown by structures **22-31.**

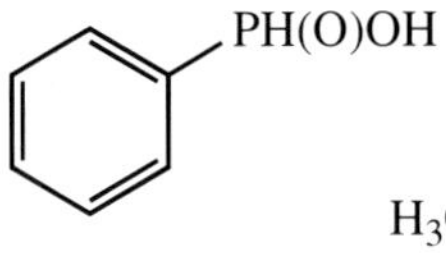

22[17, 49]

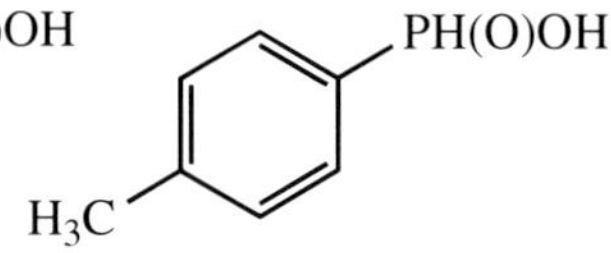

23[57]

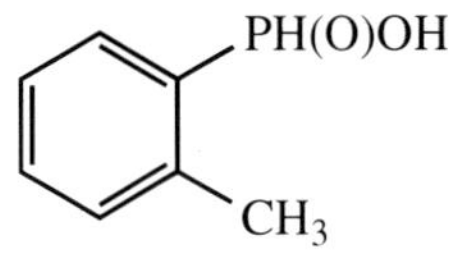

24[57]

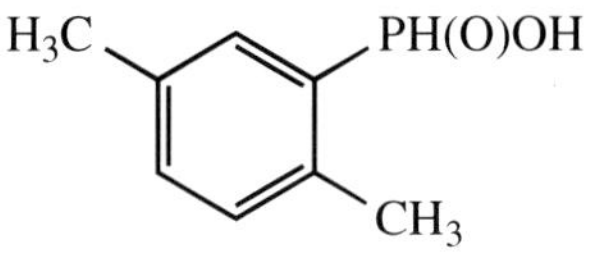

25[57]

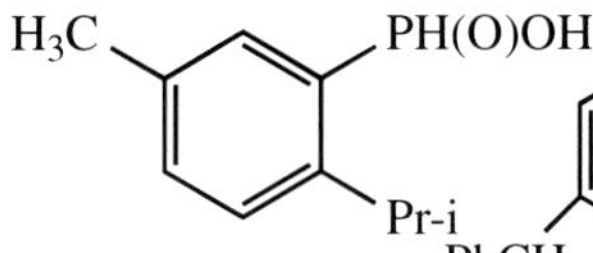

26[57]

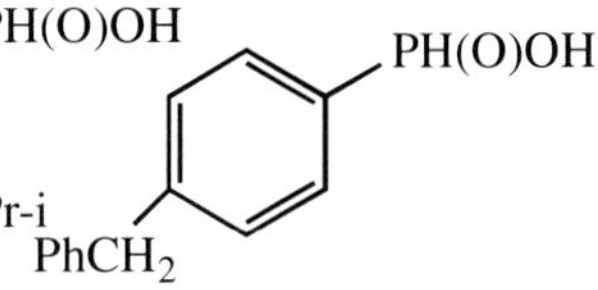

27[40]

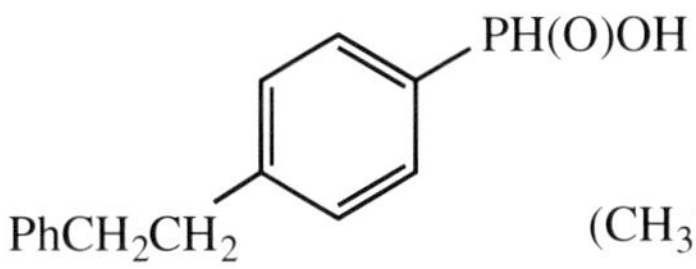

28[40]

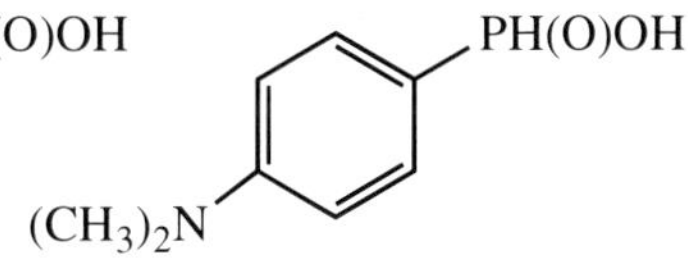

29[14a]

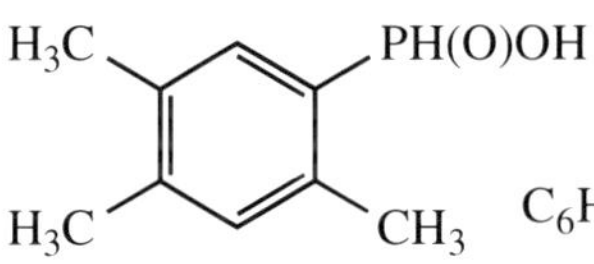

30[13

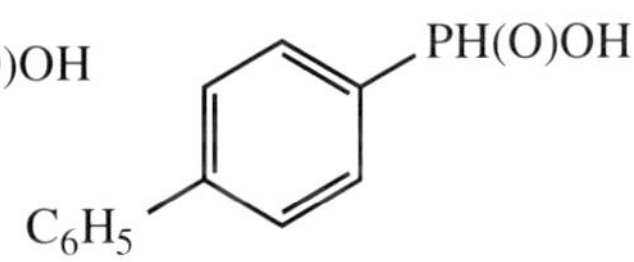

31[40]

7.D. Phosphinous Acids (Ar_2P-OH)

With the exception of one or two special cases, these are known today only as ester or amide derivatives; as in the phosphonous acids, hydrogen is attached to phosphorus and is not acidic. The compounds thus have the structure $Ar_2P(O)H$ and are properly classified as secondary phosphine oxides. This point is discussed further in Section 3.9.c. Michaelis first encountered this structural type when he performed hydrolysis on his diphenylphosphinous chloride [26] and obtained an oily product; he expressed the structure as Ph_2POH and did not comment on the lack of acidity or the location of the hydrogen. He did observe that heating the compound gave two products (see Section 3.9; a result of a disproportionation reaction), diphenylphosphine and diphenylphosphinic acid, and that oxidation also gave this acid. In 1885, he reported the reaction of Ph_2PCl with sodium phenoxide which gave the phosphinite Ph_2P-OPh [32]. He found this ester to be oxidized to the phosphinate $Ph_2P(O)OPh$. It also added bromine to give a stable adduct of structure $Ph_2PBr_2(OPh)$, and it added sulfur or selenium to give the thio- or selenophosphinic acids, Ph_2P(S or Se)(OPh). An important observation was that the ester could be alkylated with $PhCH_2Cl$ to give the salt $Ph_2P(CH_2Ph)(OPh)^+ Cl^-$ and that basic hydrolysis displaced the phenoxy group and gave the tertiary phosphine oxide $Ph_2P(O)CH_2Ph$. The alkylation of trivalent phosphorus esters is the basis for the famous Michaelis-Arbuzov reaction which is discussed in Section 3.8.

Michaelis obtained another important result in his studies on phosphinous chlorides and their hydrolysis products; he appears to have made the first heterocyclic compound with P and N in the ring, where phosphorus is a member of the 5,10-dihydrophenophosphazine ring system (Eq. 17) [14a]. This was accomplished by the reaction of diphenylamine and PCl_3 in a sealed tube at 250° to give a presumed intermediate phosphinous chloride. As for simpler phosphinous chlorides, the chlorine could be replaced by hydroxyl on hydrolysis. Michaelis wrote the structure **32** for the hydrolysis product, but years later this compound was shown to have the secondary phosphine oxide structure **33** [58].

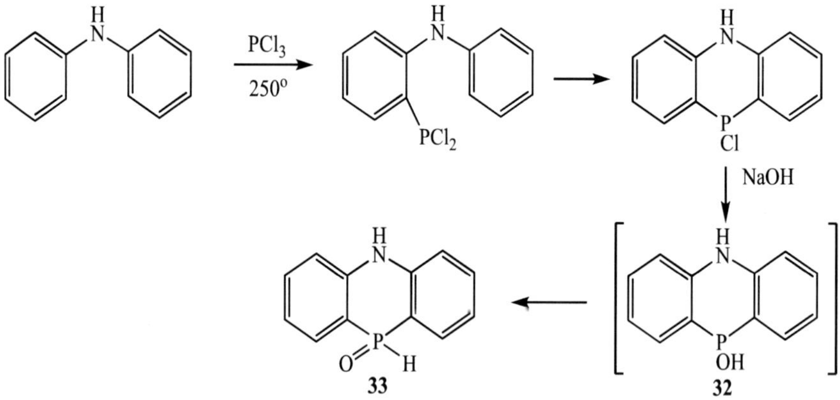

(17)

3.8. THE SYNTHESIS OF P-ALKYL PHOSPHONIC AND PHOSPHINIC ACIDS

It was previously mentioned (Section 3.5) that Michaelis described a method for preparing ethylphosphonic acid in his studies on ethylphosphonous dichloride [34] and that Guichard later prepared several other alkylphosphonic acids using his method [35]. Among Michaelis's later accomplishments was the discovery of two methods for the synthesis of alkylphosphonic acids that have far greater generality and utility than his earlier method. These methods remain today as major sources of these acids, but curiously both were presented as brief observations in papers concerned also with other matters, and not reported again in any of his later papers. Nevertheless, he was the first to describe these reactions that still bear his name today. These are the Michaelis-Becker (more commonly just Michaelis) reaction and the Michaelis-Arbuzov, also known simply as the Arbuzov reaction or rearrangement because of the extensive development of the method by A.E. Arbuzov in later years [59].

The Michaelis-Becker reaction can be expressed by Eq. 18, wherein a dialkyl phosphite (now better described as dialkyl H-phosphonate) is converted to a metallic derivative which is then alkylated on phosphorus by an alkyl halide. The original Michaelis paper employed diethyl phosphite and ethyl iodide as reactants, thus forming diethyl ethylphosphonate. [60].

$$(C_2H_5O)_2\overset{\overset{O}{\|}}{P}H \xrightarrow{Na} (C_2H_5O)_2\overset{\overset{O}{\|}}{P}^- \, Na^+ \xrightarrow{C_2H_5I} (C_2H_5O)_2\overset{\overset{O}{\|}}{P}-CH_2CH_5 \tag{18}$$

That alkylation had occurred on P and not on O was established by preparing the same compound by reaction of the known ethylphosphonic dichloride with ethanol or sodium ethoxide (Eqn. 19).

$$C_2H_5PCl_2 \xrightarrow{(O)} C_2H_5P(O)Cl_2 \xrightarrow{NaOC_2H_5} C_2H_5P(O)(OC_2H_5)_2 \tag{19}$$

That the method has been used extensively over the years is evident from the many examples cited in the book by Savignac and Iorga [61], revealing numerous modifications allowing the synthesis of functionally-substituted phosphonates.

An example of the Michaelis-Arbuzov reaction is given in Equation 20; the process also is included in the book by Savignac and Iorga [61] as well as in *Chemical Reviews* [59]. Experimental details are given in *Organic Syntheses* [62]. The efficiency of the process is evident from the yield of 85-90% reported therein for the reaction of methyl iodide and tri-isopropyl phosphite, and 98.5% for ethyl iodide and triethyl phosphite. In the special case where R' is the same as R (as in Eq. 20), the overall process amounts to a rearrangement, and has become known as the Arbuzov rearrangement.

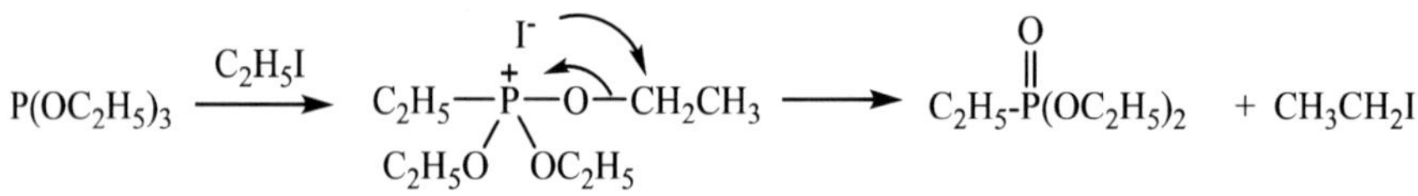

34

(20)

The original paper by Michaelis and Kaehne [63] was concerned mostly with the reaction of triaryl phosphites (which included triphenyl, tripseudocumyl, tri-p-cresyl, tri-m-cresyl and tri-p-chlorophenyl, collectively referred to as "O-phosphines") with the alkyl halides methyl iodide and benzyl chloride. The initial products were found to be crystalline salts from alkylation on phosphorus. (Eq. 21). On hydrolysis with aqueous NaOH, these were converted to diaryl esters of an alkylphosphonic acid.

$$(ArO)_3P \xrightarrow{CH_3I} (ArO)_3\overset{+}{P}-CH_3 \;\; I^- \xrightarrow{NaOH} (ArO)_2\overset{O}{\overset{\|}{P}}\text{-}CH_3 \qquad (21)$$

The alkylations with methyl iodide and benzyl chloride were then applied to triethyl phosphite, whereupon no crystalline adduct could be isolated. Instead the observed products were diethyl methylphosphonate and diethyl benzylphosphonate. These could be hydrolyzed by NaOH, followed by acidification, to give the phosphonic acids. Methylphosphonic was first prepared by A. W. Hofmann in 1872 [29] by oxidation of methylphosphine and the two compounds had comparable melting points. Michaelis speculated that the first product was the salt (**34**, Eq. 20) which then decomposed to the phosphonate. We now know this to be the case, with the decomposition described as nucleophilic attack of iodide ion on the carbon of an alkoxy group. The stability of the salts from triaryl phosphites is of course due to the reluctance of the aryloxy group to undergo attack on the ring carbon by nucleophiles. Thus in the Michaelis-Kaehne paper we have two new methods for the synthesis of alkylphosphonates, although there has been little follow-up to the triaryl phosphite method.

Another P-alkylation method was reported to give rise to arylalkylphosphinic acids [22] resulting from studies on diamino derivatives prepared from phenylphosphonous dichloride (Eq. 22). In particular, the product **35** from piperidine was reacted with methyl iodide to give the crystalline salt **36**. The iodide ion was replaced with hydroxide ion by treatment with moist silver oxide; hydrolysis then gave phenylmethylphosphinic acid. p-Tolylmethylphosphinic acid was prepared similarly.

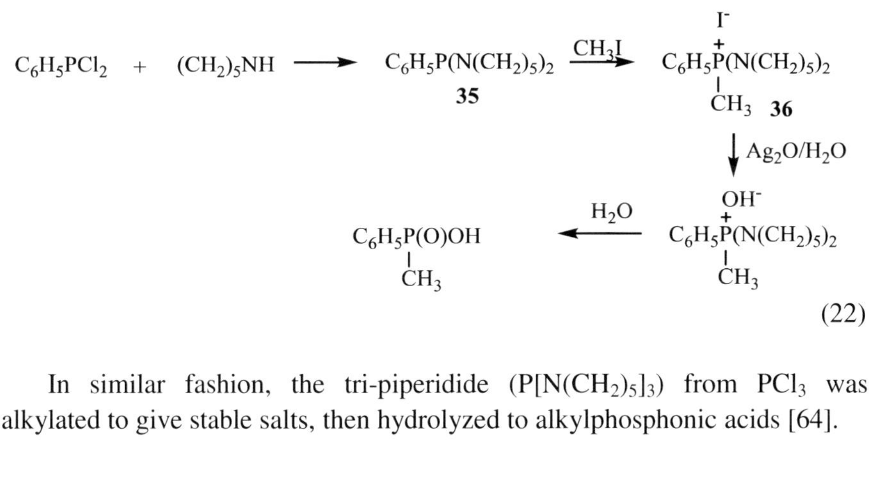

(22)

In similar fashion, the tri-piperidide ($P[N(CH_2)_5]_3$) from PCl_3 was alkylated to give stable salts, then hydrolyzed to alkylphosphonic acids [64].

3.9. Other Investigations

9.A. Reactions Making P-S Bonds

Several attempts were made to prepare compounds with P-S bonds, most with uncertain results or structures that are unlikely and need confirmation. The best result came from extending the use of phosphorus halides in Friedel-Crafts type reactions to include thiophosphoryl chloride ($PSCl_3$). This compound reacted with toluene in the presence of aluminum chloride, thus providing a new way to attach phosphorus to the aromatic ring. Both mono-aryl and di-aryl compounds were obtained from the reaction with toluene (Eq.23), with the ditolyl compound predominating. Reactions with ammonia or amines, or with alcohols, gave the expected substitution products (**37** and **38**, respectively) [40]. Michaelis only reported the reaction with toluene, and did not develop this process further.

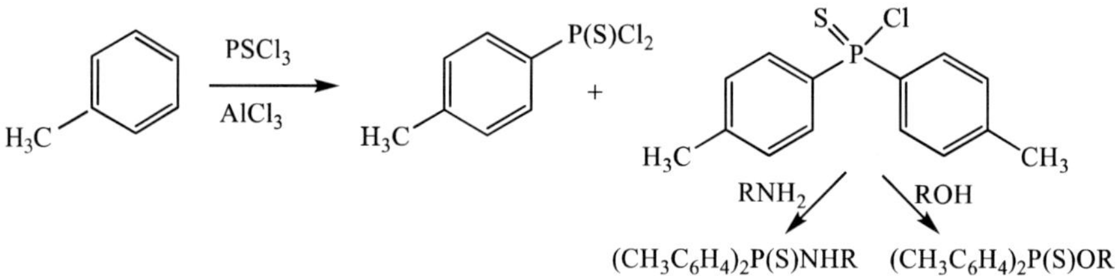

37 **38**

Scheme 10.

Other experiments on reactions of phosphines and CS_2 [65] and of phosphonous dichlorides with hydrogen sulfide [66] gave products whose structures were difficult to elucidate at that time, and remain unclear. They will not be further considered in this paper.

9.B. Reaction of Trivalent Phosphorus Halides with Acetone

Michaelis published two papers on the reaction of PCl_3 and $AlCl_3$ with acetone, and was able to prove that after hydrolysis he had obtained a phosphonic acid [67, 68] with proposed structure **39.**

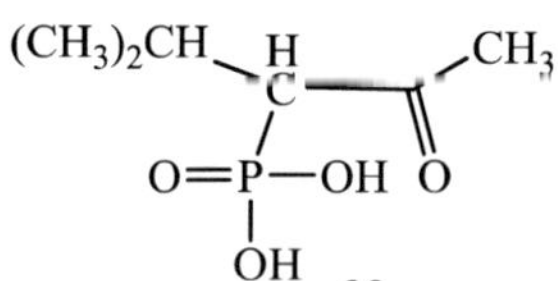

39

We now know this structure to need revision, according to work of Drake and Marvel [69] and of Anschütz, Klein and Cermak [70]. The reaction is presumed to proceed first through the formation of mesityl oxide from aldol condensations of acetone, and then 1,4-addition of PCl_3 to the conjugated system to give **40** (Scheme 11). The structure proposed for the hydrolysis product is shown as **41**, differing from that of Michaelis by the position of the phosphonic acid group. Nevertheless, Michaelis deserves credit for successfully showing that a new type of phosphonic acid with a keto function can be prepared from acetone.

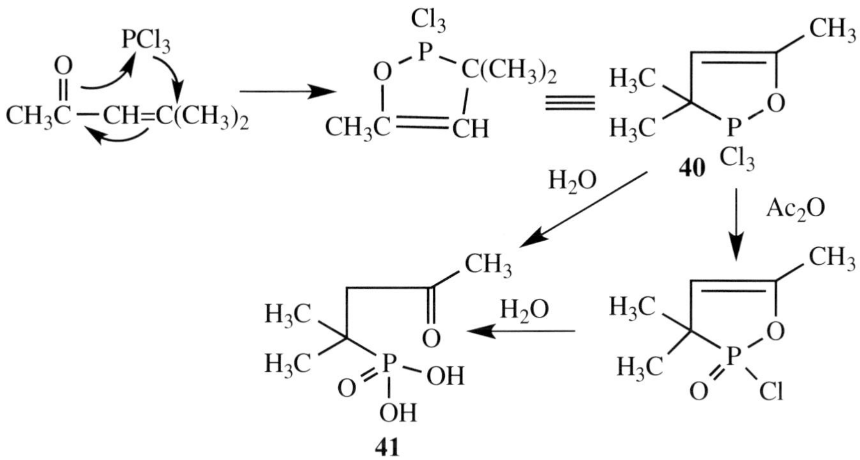

Scheme 11.

Michaelis later used $C_6H_5PCl_2$ in the reaction with acetone and obtained a phosphinic acid to which he assigned a related structure **42** [71]. More likely the true structure is **43.**

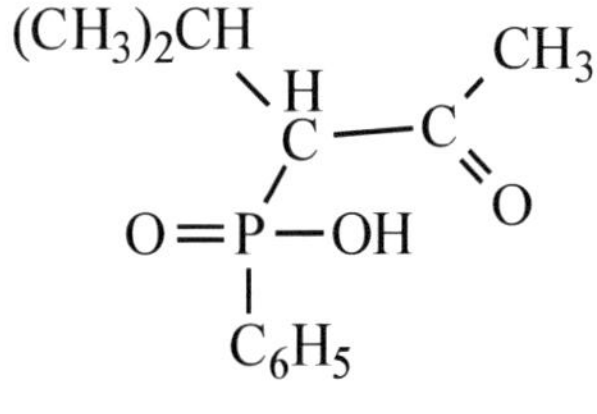

42

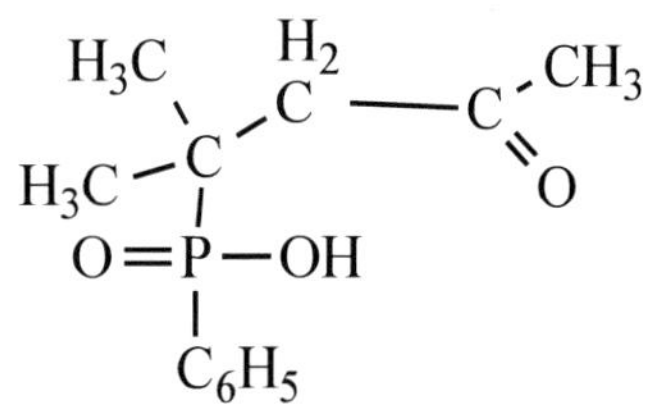

43

9.C. Observations on the Structure of Trivalent P-Hydroxy Compounds

We now know that in general trivalent P-hydroxy compounds are involved in a form of tautomeric equilibrium where one hydrogen shifts to the P atom; this tautomer always is in great predominance, with few exceptions.

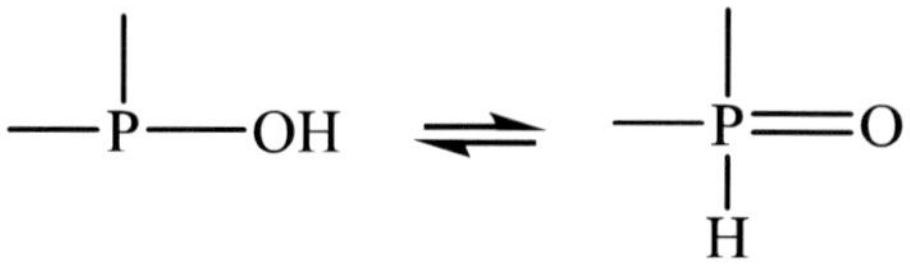

(23)

Michaelis became aware of this type of shift from studies on phosphorous acid, which he properly expressed as $(HO)_2PHO$ rather than as $P(OH)_3$ since it is dibasic [19]. He also wrote phenylphosphonous acid correctly as PhPH(O)(OH) rather than $PhP(OH)_2$ because it is monobasic. Nevertheless, he noted that both of these acids could be converted with PCl_5 to the trivalent chlorides PCl_3 and $PhPCl_2$, respectively, which left some ambiguity about the proper structure since these chlorides seemed to be coming from $P(OH)_3$ and $PhP(OH)_2$. In this 98-page article, Michaelis reviews the work of others, especially A.W. Hofmann, and his own work. He later [62] returned to the P(OH) structural problem and confirmed that one hydrogen was indeed different in the acids and must be bonded to phosphorus. He noted that the hydrogen of dialkyl phosphites could be removed with metals (notably sodium) and the anion alkylated on P by alkyl halides. As seen in Section 3.8, this reaction has been developed over the years into a highly useful procedure for the synthesis of alkylphosphonic esters and acids, and he and Becker have been honored by the naming of the reaction as the Michaelis-Becker Reaction. As noted in Section 3.7, Michaelis simply wrote the structure "Ph_2POH" for the product of hydrolysis of Ph_2PCl and of Ph_2PNHPh [26]. The proton rearrangement would give the structure $Ph_2PH(O)$, diphenylphosphine oxide (**44**). He referred to it as a "phosphinigesäure" (phosphinous acid), and found it to be unstable through a disproportionation to diphenylphosphine and diphenylphosphinic acid (Scheme 12). However the product was not further characterized. Diphenylphosphine oxide was not isolated as a pure crystalline compound until 1957 [72], when its true phosphine oxide structure was established.

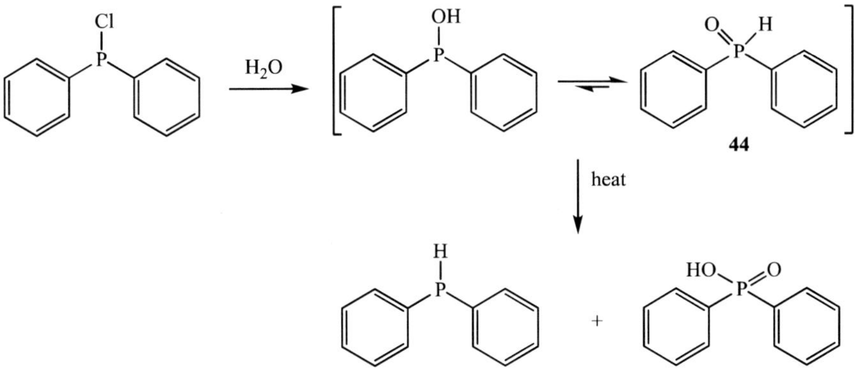

Scheme 12.

These early observations have been extended to many other compounds over the years and it is a rule of phosphorus chemistry, with only one or two exceptions [20a], that stable compounds with OH on trivalent phosphorus cannot exist and are properly treated as the product of proton migration from OH to the P atom. Michaelis's insight and work did much to clarify this property of P(OH) compounds.

9.D. Observations on the Valence of Phosphorus

It must be remembered that Michaelis conducted much of his research when the powerful Structural Theory of Kekulé and others (see Chapter 1) had only been propounded a few years earlier. As early as 1872, Michaelis recognized that phosphorus could have a valence of 3 or 5 [73]. In an 1885 paper [74] he discussed proper structural formulas to use for compounds that can be generalized as addition products of trivalent phosphorus, such as PF_5 and PCl_5. Thus, did P retain its trivalency and had the other atoms been held as some sort of complex, or did phosphorus actually have a valence of five? This paper uses the expression $PX_3{\cdot}X_2$, but we now know that P can indeed have five groups covalently bonded (although in some media the proper formulation for halides is ionic, as in is PX_4^+ PX_6^-). And what to do about two isomeric compounds, totally different in their characteristics, but both possessing the molecular formula $C_{18}H_{15}OP$? These of course are his triphenylphosphine oxide and phenyl diphenylphosphinite. He concluded that their structures must be **45** and **46**, respectively, drawn just as he did, and are thus pentavalent and trivalent.

$(C_6H_5)_3P{\equiv}O$ $(C_6H_5)_2P$---O---C_6H_5

45 **46**

Structure **45** is an early example of an expression of structure for the phosphoryl group (P=O), written with a double bond between P and O to show phosphorus as pentavalent, but not to be confused with the common *pi* double bond (e.g., C=C, C=N, C=O, etc.) that is derived from two sets of orbital interactions. The expression P=O we still use today, but it does not represent the true nature of this bonding, which has been and remains the subject of much theoretical consideration and debate (see for example, reference 75).

CONCLUSION

August Michaelis performed a massive amount of research in organophosphorus chemistry, which is summarized in this chapter. He discovered several reactions that are still important in modern phosphorus chemistry, and he was the first to observe and characterize a number of new functional groups containing phosphorus. He was the dominant figure in organophosphorus chemistry from 1873 to the time of his death in 1916.

REFERENCES

[1] Michaelis, A. *Ber.* 1872, *5,* 4, 6, 9*;* see also [73].
[2] Michaelis, A. *Ber.* 1873, *6,* (a) 601; (b) 816. See also Michaelis, A.; Mathias, C. *Ber.* 1874, *7*, 1070.
[3] Wichelhaus, H. *Ber.* 1916, *49,* 468.
[4] Hofmann, A. W. *Ber.* 1872, *5,* 100.
[5] (a) Toy, A. D. F.; Walsh, E. N. *Phosphorus Chemistry in Everyday Life*; American Chemical Society: Washington, DC, 1987; p 228. (b) Abramson, A. A.; Telschow, J. E. U.S. Patent 5698736A, Dec. 16, 1997.
[6] Michaelis, A. *Ann.* 1876, *181,* 265.
[7] Michaelis, A.; Lange, H. *Ber.* 1875, *8*, 1313.
[8] Michaelis, A. *Ber.* 1879, *12*, 1009.

[9] Kosolapoff, G. M. *Organophosphorus Compounds*; John Wiley: New York, 1950, p 54.
[10] Michaelis, A.; Panek, C. *Ber.* 1880, *13*, 653.
[11] *Organic Syntheses*, Coll. Vol. *4,* 1963, p 784.
[12] Dye, W. T., Jr. *J. Am. Chem. Soc*. 1948, *70*, 2595.
[13] Michaelis, A. *Ann.* 1897, *294,* 1.
[14] (a) Schenk, A.; Michaelis, A. *Ber.* 1888, *21*, 1497; (b) Michaelis, A.; Schenk, A. *Ann.* 1890, *260,* 133, 1808.
[15] Bourneuf, M. *Bull. Soc. Chim. Fr.* 1923, *33,* 1808.
[16] Michaelis, A; Köhler, H. *Ber.* 1876, *9*, 519.
[17] Michaelis, A.; Ananoff, J. *Ber.* 1874, 7, 1688.
[18] (a) Michaelis, A. *Ber.* 1874, *7*, 6. (b) Köhler, H.; Michaelis, A. *Ber.* 1877, *10,* 807.
[19] Michaelis, A. *Ber.* 1875, *8*, 504.
[20] Quin, L. D. *A Guide to Organophosphorus Chemistry:* John Wiley; New York, 2000, (a) pp. 20-21, (b) p. 8.
[21] Köhler, H.; Michaelis, A. *Ber.* 1877, *10*, 816.
[22] Michaelis, A. *Ber.* 1898, *31*, 1037.
[23] Köhler, H; Michaelis, A. *Ber.* 1876, *9*, 1053.
[24] Cahours, A.; Hofmann, A. W. *Ann.* 1857, *104*, 1.
[25] Michaelis, A. *Ber.* 1875, *8,* 493.
[26] Michaelis, A.; Gleichmann, L. *Ber.* 1882, *15,* 801.
[27] (a) Michaelis, A.; Reese, A. *Ber.* 1882, *15,* 1610. (b) Michaelis, A.; v. Soden, H. *Ann.* 1885, *229*, 295.
[28] Michaelis, A.; Graeff, F. *Ber.* 1875, *8*, 922, 1304.
[29] Hofmann, A. *Ber.* 1873, *6,* 303.
[30] Michaelis, A. *Ber.* 1877, *10*, 627.
[31] Broglie, A. *Ber.* 1877, *10*, 628,
[32] Michaelis, A.; La Coste, W. *Ber.* 1885, *18*, 2109.
[33] Michaelis, A.; Link, A. *Ann.* 1881, *207*, 193.
[34] Michaelis, A. *Ber.* 1880, *13*, 2174.
[35] Guichard, F. *Ber.* 1899, *32,* 1572.
[36] (a) Hofmann, F. W.; Moore, T. B. *J. Am. Chem. Soc.* 1958, *80,* 1150. (b) Pianfetti, J. A.; Quin, L. D. *J. Am. Chem. Soc.* 1962, *84*, 851.
[37] Regitz, M.; Scherer, O. J. *Multiple Bonds and Low Coordination in Phosphorus Chemistry;* Georg Thieme: Stuttgart, 1990; (a) p 415, (b) p 293.
[38] Michaelis, A. *Ann.* 1903, *326,* 129.

[39] Michaelis, A. *Ann.* 1896, *293*, 193 (article as published is incomplete; data taken from Ref. 9, p 147); see also [17].
[40] Michaelis, A. *Ann.* 1901, *315*, 43.
[41] Michaelis, A.; Flemming, A. *Ber.* 1901, *34*, 1291.
[42] Michaelis, A.; Rothe, F. *Ber.* 1892, *25*, 1747.
[43] (a) Michaelis, A.; Schroetter, G. *Ber.* 1894, *27*, 490. (b) Michaelis, A.; Oster, F. *Ann.* 1892, *270*, 123.
[44] Michaelis, A.; v. Soden, H. *Ber.* 1884, *17*, 921.
[45] (a) Hofmann, A. W. *Proc. Roy. Soc.* 1862, 11, 25. (b) Meyer, A. H. *Ber.* 1871, *4,* 734.
[46] Michaelis, A.; Gimborn, H. *Ber.* 1894, *27*, 272.
[47] Wittig, G.; Rieber, M. *Naturwissenschaften* 1947, *35,* 345.
[48] Yoshifuji, M.; Shima, N.; Inamoto, N.; Hirotsu, K.; Higuchi, T. *J. Am. Chem. Soc.* 1981, *103*, 4587.
[49] Götter, H.; Michaelis, A. *Ber.* 1878, *11,* 885.
[50] Kuchen, W.; Buchwald, H. *Chem. Ber.* 1958, *91*, 2296, and papers cited in Daly, J. J. *J. Chem. Soc.* 1964, 6147.
[51] Hofmann, A. *Ber.* 1872, *5*, 104.
[52] (a) Michaelis, A. *Ann.* 1896, *293,* 261. (b) a related paper is Michaelis, A.; Kammerer, F. *Ber.* 1875, *8,* 1306.; (c) Michaelis, A. *Ann.* 1915, *407*, 290.
[53] Michaelis, A.; Benzinger, E. *Ann.* 1877, *188,* 275.
[54] Michaelis, A.; Benzinger, E. *Ber.* 1875, *8*, 1310.
[55] Michaelis, A.; Panek, C. *Ber.* 1881,*14*, 405.
[56] Michaelis, A.; Wegner, F. *Ber.* 1915, *48*, 316.
[57] Michaelis, A.; Panek, C. *Ann.* 1882, *212*, 203.
[58] Häring, M., *Helv. Chim. Acta* 1960, *43*, 1826. See also Freedman, L.D.; Freeman, H. S. *Chem. Rev.* 1987, *87*, 289.
[59] Bhattacharya, A. K.; Thyagarajan, G. *Chem. Rev.* 1981, *81*, 415.
[60] Michaelis, A.; Becker, T. *Ber.* 1897, *30*, 1003.
[61] Savignac, P; Iorga, B. *Modern Phosphonate Chemistry*: CRC Press; Boca Raton, FL, 2003.
[62] *Organic Syntheses,* Coll. Vol. *4*, 1963, 325.
[63] Michaelis, A.; Kaehne, R. *Ber.* 1898, *31*, 1048.
[64] Michaelis, A.; Luxembourg, K. *Ber.* 1896, *29,* 2205.
[65] Michaelis, A.; Dittler, F. *Ber.* 1879, *12*, 338.
[66] Köhler, H.; Michaelis, A. *Ber.* 1877, *10*, 815.
[67] Michaelis, A. *Ber.* 1884, *17,* 1273.
[68] Michaelis, A. *Ber.* 1885, *18,* 898.

[69] Drake, L. R.; Marvel, C. S. *J. Org. Chem.* 1937, *2*, 387.
[70] Anschütz, L.; Klein, E.; Cermak, G. *Ber.* 1945, *77*, 726.
[71] Michaelis, A. *Ber.* 1886, *19,* 1009.
[72] Hunt, B. B.; Saunders, B. C. *J. Chem. Soc.* 1957, 2413.
[73] Michaelis, A. *Ann.* 1872, *164,* 9.
[74] Michaelis, A.; La Coste, W. *Ber.* 1885, *18,* 2118.
[75] Gilheany, D. G. *Chem. Rev.* 1994, *94,* 1339.

Chapter 4

THE CONTRIBUTIONS OF OTHER CHEMISTS, 1870-1900

ABSTRACT

Many chemists entered the field of organophosphorus chemistry as they became aware of the original work of Hofmann and Michaelis. Their research efforts added new synthetic methods and novel compounds in the areas of phosphorus acids, phosphines, salts and oxides of phosphines, and phosphorus halides. The research publications of these chemists, estimated to be about 45 in number, are reviewed in this Chapter.

4.1. INTRODUCTION

Before Michaelis commenced his monumental program in phosphorus chemistry, only one other chemist in the Hofmann era had published a paper in the field. L. Pebal in Lemberg University in 1861 [1] contributed a novel idea that was later developed with the use of different reagents. Pebal devised a synthesis of triethylphosphine oxide by reacting phosphorus oxychloride ($POCl_3$) with diethylzinc. His procedure led to an explosively fast reaction, but did give the desired product. Zinc reagents fell out of use when Grignard reagents were later introduced, and now the oxychlorides of phosphorus are useful starting materials for the synthesis of the carbon-phosphorus bond.

From the date of his first publication on organic phosphorus compounds in 1873, August Michaelis dominated the field of organophosphorus chemistry for the rest of the nineteenth century. However, contributions to the field by other chemists began to appear during this period, especially from 1880 to 1900, as they recognized the novelty of this new area. Some of this work was performed in the institutes where Michaelis had his laboratories (Aachen and Rostock), and in a few papers the comment was made that Professor Michaelis had proposed or had consulted on the project. Research on phosphorus chemistry continued to develop in Germany, but significant contributions were made during the latter part of the nineteenth century by chemists in France and Great Britain. It was not until the early twentieth century that research on phosphorus chemistry became of importance in Russia, where Alexander E. Arbuzov became the leading figure in this emerging field.

In reviewing the research of the various contributors, it is convenient to do so by considering the advances made with specific phosphorus functional groups, rather than by investigator as was done with Hofmann and Michaelis, since many of the contributions were made in a single publication with no strong continuing program emerging. Some of the more significant contributions were made in the area of phosphorus acids, which will be discussed first. Phosphines will be considered next, followed by the phosphine derivatives (phosphonium salts and phosphine oxides). Finally the organic phosphorus halides will be reviewed. A difficulty in this approach is that there is considerable overlap in these areas; for example, some of the new results in acids chemistry depend upon new results on the synthesis of phosphorus halides.

4.2. Phosphorus Acids

4.A. New Synthetic Methods

Methods still valuable today were developed using aldehydes as starting materials. The first of these was reported by Wilhelm Fossek in a preliminary report in 1884 [2] on the addition of phosphorus trichloride to aldehydes (Eq. 1). The product gave crystalline compounds on hydrolysis, but he drew no structures for these products. In a longer paper [3] he showed that the hydrolysis products contained no chlorine, and he correctly drew the structures as alpha-hydroxyalkylphosphonic acids, $RCH(OH)P(O)(OH)_2$. He applied this reaction to propionaldehyde, isobutyraldehyde, isovaleraldehyde, and benzaldehyde. He continued to work in this area [4] and described the isolation by distillation of the initial addition product (**1**) which gave the correct elemental analysis. The acid chlorides reacted with water to give the Cl-free phosphonic acid **2,** and with ethanol to give phosphonate esters with retention of the C-Cl group. An important discovery was that the OH group could be reduced by HI, thus giving unsubstituted alkylphosphonic acids (Eq. 1). At that time, the only other synthesis of such acids was the Hofmann oxidation of primary phosphines [5].

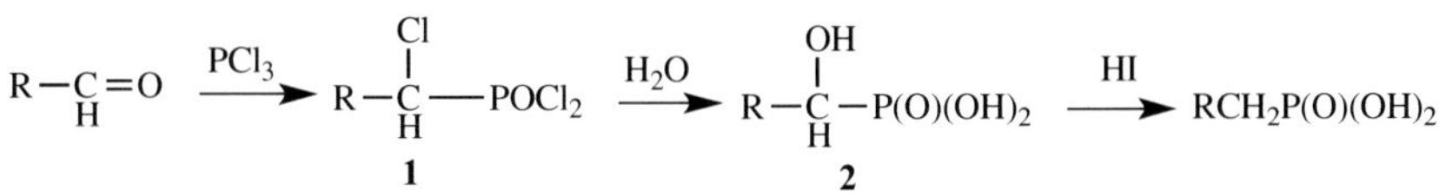

(1)

A different phosphorus reagent, hypophosphorous acid (**3**), was used by J. Ville for the synthesis of hydroxyalkylphosphonous acids [6]. As seen in Eq. 2, the initial 1:1 adduct (**4**), still retaining a reactive P-H bond, could react to give a di(hydroxyalkyl)phosphinic acid (**5**), but control of the reactant ratio was effective in giving the mono-alkylphosphonic acid [7].

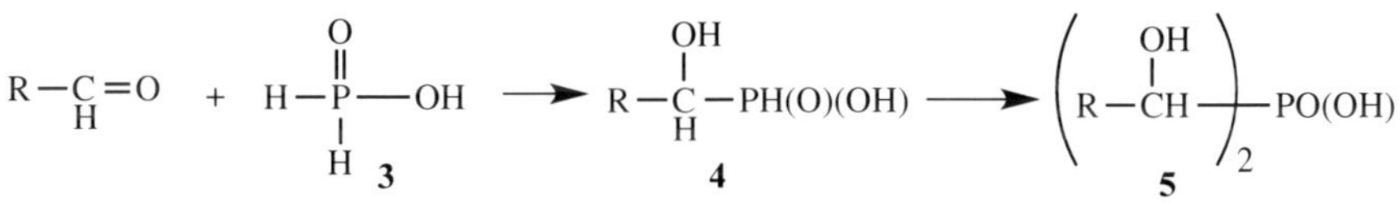

(2)

Ville found [8,9] that the reaction was quite general and applied it to a variety of aldehydes, both aromatic and aliphatic. He also provided a method for preparing simple esters from the acid, involving the formation of the silver salt and its alkylation by alkyl iodides.

Benzaldehyde was used in a quite different reaction reported by S. Litthauer [10]. This led to benzylphosphonic acid (**6**), as well as to dibenzylphosphinic acid (**8**) and tribenzylphosphine oxide. Relative amounts were not given. The phosphorus reactant was phosphonium iodide (PH_4I); the reaction was conducted at 100°. It was presumed by Litthauer that the phosphonium ion reduced the carbonyl group, forming benzyl alcohol, which then alkylated the phosphorus atom. However, Kosolapoff [11] suggests that the usual P-H addition to the carbonyl group occurs, and the C-hydroxy product is reduced by the phosphonium ion. Litthauer performed nitration with cold fuming nitric acid on both of his acids **6** and **8**, and obtained the mononitrobenzyl products (**7**) and (8), respectively (Eq. 3 and 4). He did not report the position of the nitro group but it is undoubtedly in the para-position as shown.

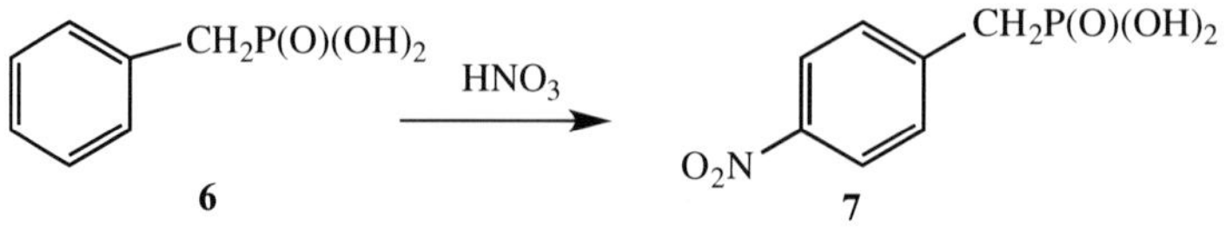

(3)

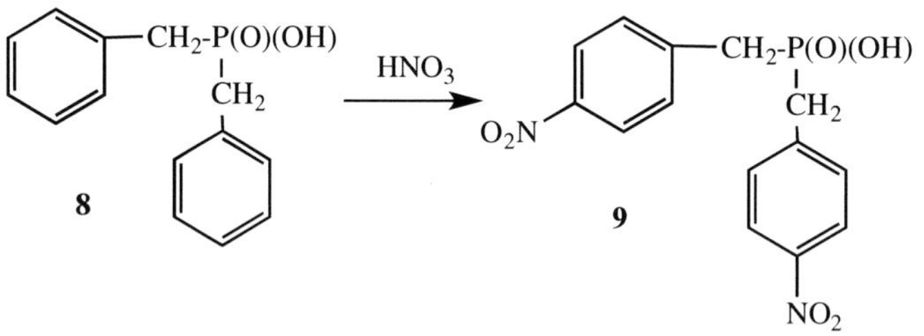

(4)

In the same year, Letts prepared benzylphosphonic acid by the Hofmann route of oxidizing benzylphosphine [12]. Neither method finds use today, but they were of value in adding an important member to the list of phosphonic acids discovered in the nineteenth century.

There are several papers in the late nineteenth century that described reaction products with structures that are questionable or wrong. These are included here to help make the point that phosphorus research was spreading as the century approached its end.

(1) Another carbonyl compound, acetophenone, figured in the 1880's burst of activity in phosphonic acid chemistry. In studying the reaction of acetophenone with PCl_5, A. Behal [13] observed not only the expected dichloro product from replacement of the carbonyl oxygen, but also the formation of a phosphonic acid, assigned structure $PhCOCCl_2P(O)(OH)_2$. Only an analysis was given to suggest this structure, but it seems unlikely that a C-P bond would form from PCl_5 in such a way.

(2) Another reaction with a carbonyl compound is that described by A. Claus [14], where he reacted PCl_5 with the amides acetyl- and benzoyl-diphenylamine. After hydrolysis, the product from the former was given formula **10** as one possibility, and shown also as **11** by Fossek in one of his papers [4]. However, this structure too is unconfirmed, again because C-P bond formation from PCl_5 in this manner is unlikely.

$Ph_2N\text{--}C_2H_3.Cl.(PO_3H_2)$

10

$$H_3C-\overset{\displaystyle Cl}{\underset{\displaystyle NPh_2}{C}}-PO_3H_2$$

11

(3) Bishop, Claisen and Sinclair [15] reacted a compound they called oxymethylene camphor, shown by them as structure **12**, with PCl_3 or PBr_3 and obtained a product $C_{11}H_{17}O_4P$. They considered this to be a phosphite of structure **13**. However, Michaelis and Flemming [16] reported that the phosphorus was not removed by hydrolysis and the compound must be an isomeric hydroxyphosphonic acid (**15**). This is easily explained by assuming the enol **12** reacts as the aldehyde tautomer, which then adds PCl_3 to form the trichloride **14**. This is the Fossek process. Hydrolysis gives the Fossek hydroxyphosphonic acid **15**, and Michaelis and Flemming prepared some derivatives of this acid. Dehydration of hydroxy acid **15** would give the Bishop product, but the pathway remains uncertain.

$(C_{10}H_{14}O)\text{=}CH\text{-}OH$ **12**

$(C_{10}H_{14}O)\text{=}CH\text{-}O\text{-}P(OH)_2$ **13**

$(C_{10}H_{14}O)\text{-}\overset{Cl}{C}HP(O)Cl_2$ **14**

$(C_{10}H_{14}O)\text{-}\overset{OH}{C}HP(O)(OH)_2$ **15**

(4) In research aimed at determining structural relations of the terpenes camphene and fenchene, and of camphor and fenchone, phosphorus pentachloride was used in reactions with these compounds. Phosphonic acids were formed in some of the reactions; thus camphene, whose full structure was unknown at that time, gave a compound of formula $C_{10}H_{15}\ (PO_3H_2)$. Without adequate structural information on camphene, it was not possible to make an assignment of structure to this compound [17]. Similarly, fenchone and fenchene gave phosphonic acids of unknown structure [18,19]. A modern study of these compounds might reveal their structures, which could prove to be of some interest.

4.B. Syntheses of New Phosphonic Acids

The Michaelis reaction of aromatic compounds with PCl_3-$AlCl_3$ attracted the attention of other chemists, who made contributions of their own. The most novel of these contributions was by H. Sachs of Rostock [20], who performed the reaction on the heterocyclic aromatic compound thiophene. The general similarity in some properties between benzene and thiophene prompted this study. Indeed a phosphonous dichloride (**16**) was obtained. It boiled at 215-220° and gave the correct elemental analysis. As in other electrophilic substitutions on thiophene, the PCl_2 was placed in the alpha-position. This novel compound then gave the phosphonous acid **17** and the phosphonic acid **18**.

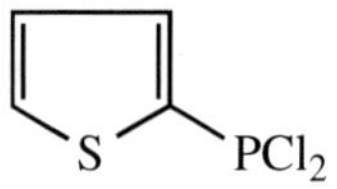

16

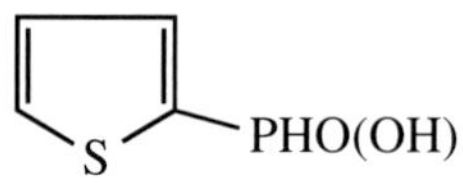

17

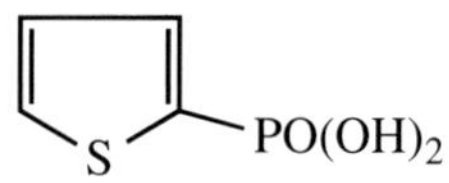

18

J. Weller of Aachen added another hydrocarbon, m-xylene, to the list of those undergoing the PCl_3-$AlCl_3$ reaction [21]. Here isomeric phosphonous dichlorides (e.g. **19**) would be formed, and were present in the distillate of the reaction product. Weller converted the mixture to phosphonic dichlorides by the familiar procedure of reaction with chlorine to form the tetrachloride, followed by sulfur dioxide, and was able to separate the mixture by distillation. The major isomer (**20**) was hydrolyzed to give the phosphonic acid (**21**). He proved the isomer structure by comparing the compound with the phosphonous dichloride prepared from 2,4-dimethylbromobenzene via the reaction of di(2,4-dimethylphenyl)mercury with PCl_3. The identity of the two products proved that the major isomer from the PCl_3-$AlCl_3$ reaction was 2,4-dimethylphosphonous dichloride (**19**), hardly a surprise since from the rules of aromatic substitution this was to be expected. (Eq. 5).

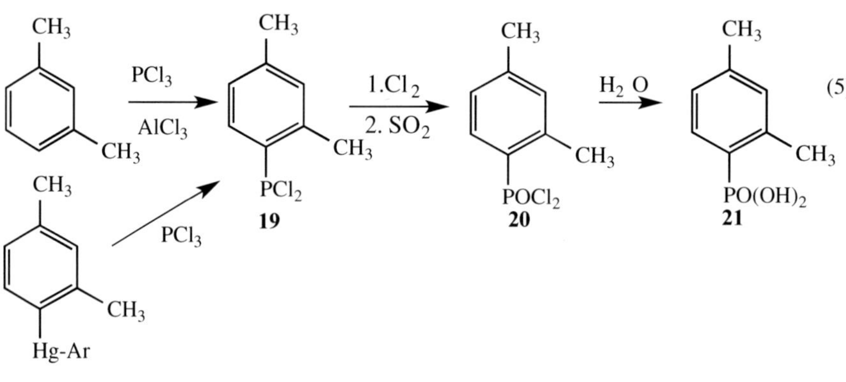

(5)

In further studies, Weller performed methyl group oxidation on phosphonic acid **21**, as well as nitration; the latter reaction gave acid **22.** In a later paper [22], Weller prepared the phosphonous dichloride from p-xylene and through the usual reactions converted this to the phosphonic acid **23.**

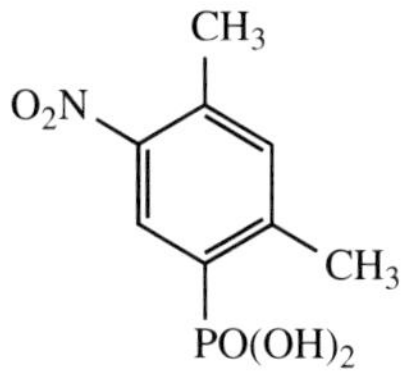

22

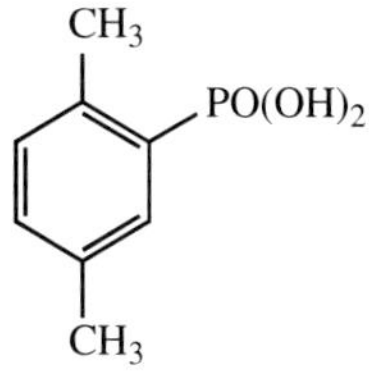

23

W. Kelbe in Karlsruhe [23] prepared the first naphthyl phosphorus compounds, employing the Michaelis method of reacting dinaphthylmercury with PCl_3 to form naphthylphosphonous dichloride. He did not state if he used 1-naphthyl or 2-naphthyl mercury derivatives, but in 1929 his products were shown to have phosphorus in the 1-position [24]. By the usual route of chlorine addition and hydrolysis of the tetrachloride, he prepared naphthylphosphonic acid. In a later paper [25] he reported the synthesis of naphthylphosphonous acid by hydrolysis of the phosphonous dichloride.

F. Melchiker, in Rostock, studied the products of the reaction of o-chlorotoluene with the PCl_3 -$AlCl_3$ mixture [26] (Eq. 6). He used a clever procedure to prove the isomer structure of the product. He first converted it to the phosphonic acid (**24**). By the action of bromine on the acid, the phosphonic group was replaced by bromine, a reaction type first reported by Michaelis [27], which gave the known 4-bromo-2-chlorotoluene (**25**). The substitution on the o-chlorotoluene therefore followed the rules of substitution, occurring para- to the ring activating methyl group.

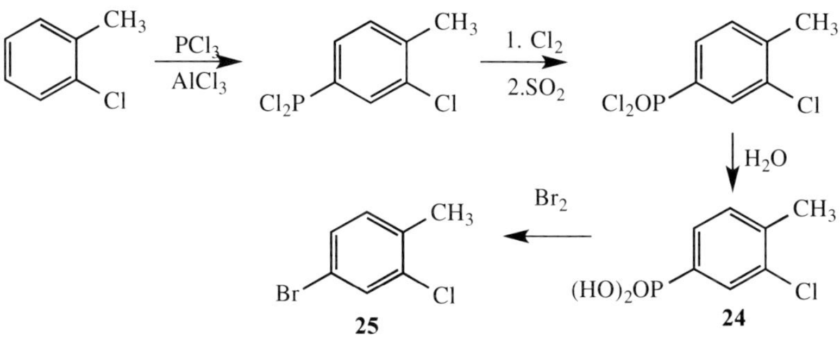

(6)

4.3. Phosphines

After the pioneering work of Hofmann on alkyl phosphines, and of Michaelis on aryl phosphines, progress on the synthesis and reactions of phosphines remained limited for the rest of the nineteenth century. As will be seen, some new phosphines were prepared by known methods, but a new synthetic method of some generality was introduced by N. Collie [28] and gave mixed tertiary phosphines. It was of some importance at the time, but less so at present, and consisted of the thermal decomposition of quaternary phosphonium chlorides that contained two or more P-ethyl groups. The decompositions were performed by pyrolytic distillation at 300°, which would be quite unattractive to modern phosphorus chemists, although the yields are good. Invariably, among the salts decomposed, ethylene was eliminated from the ethyl group, the product being the hydrochloride of a tertiary phosphine. The process is illustrated by the synthesis of isopropyldiethylphosphine hydrochloride starting with triethylphosphine (Eq. 7).

$$(C_2H_5)_3P \xrightarrow{i\text{-}C_3H_7Cl} (C_2H_5)_3(i\text{-}C_3H_7)\overset{+}{P}\ \overset{-}{Cl} \xrightarrow{heat} (C_2H_5)_2\overset{+}{P}H(i\text{-}C_3H_7)\ Cl^- + CH_2{=}CH_2 \qquad (7)$$

The new phosphines synthesized by this method included dimethylethyl-, diethylmethyl-, isopropyldiethyl, isoamyldiethyl-, benzyldiethyl- and dibenzylethyl- phosphines. In an earlier paper with E. A. Letts [29], it had been shown that tetraethylphosphonium salts with oxygen-containing anions decomposed by several pathways which produced triethylphosphine as expected but also triethylphosphine oxide.

The reaction of zinc phosphide with alkyl iodides was used by Drechsel and Finkelstein to prepare tertiary phosphines [30]. The zinc phosphide was prepared in ether solution from diethylzinc and PH_3; the alkyl iodide was added, and the mixture heated in a sealed tube at 150°. The product had the composition of the double salt $R_3P(HI)(ZnI_2)$. It was broken with aqueous KOH to provide the tertiary phosphine. Synthesized were trimethylphosphine and triethylphosphine. Using a lower temperature and a smaller amount of methyl iodide, an attempt was made to prepare monomethylphosphine. The product was isolated only as the double salt $MePH_2(HI)(AuCl_3)$.

The first member (**26**) of the new family of diphosphines ($Ar_2P\text{-}PAr_2$) was prepared in 1888 by Carl Dörken [31] using the reaction of diphenylphosphinous chloride and diphenylphosphine (Eq. 8). Dörken thus opened a whole new area; diphosphines achieved major importance in the twentieth century as ligands in coordination chemistry.

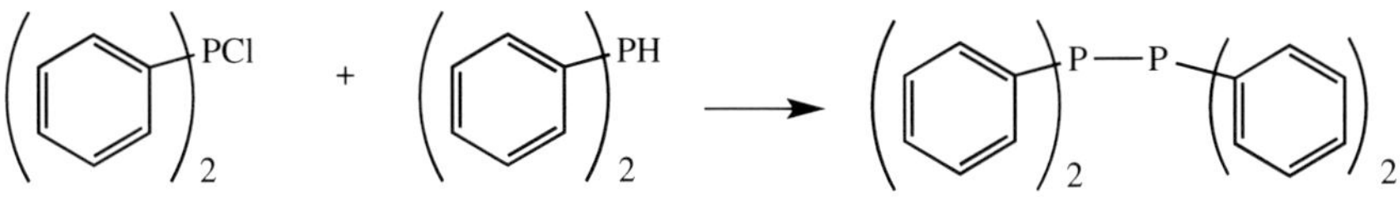

26

(8)

In this paper, Dörken also presented a synthesis and some properties of diphenylphosphine. He first reacted diphenylphosphinous chloride with metallic zinc at 230° in a sealed tube. The product gave a chlorine analysis for the composition Ph_2P-ZnCl. On hydrolysis, this material gave diphenylphosphine, which had previously been prepared by Michaelis as described in Chapter 3. He proceeded to show that the phosphine had sufficient basicity to form a hydrochloride on reaction with dry HCl. It also formed an adduct with carbon bisulfide, to which he assigned the formula $CS(PPh_2)(SPPh_2)H_2$ by C,H and S analysis, but this unlikely formulation was unsupported by other data. Dörken also made the phosphine Ph_2(p-tol)P by the Michaelis method of reacting Ph_2PCl with p-bromotoluene and sodium, and similarly he prepared Ph(p-tol)$_2$P from $PhPCl_2$.

It was noted in Section 4.2 that Sachs had prepared the novel compound thienylphosphonous dichloride (**16**) [20]. He was able to alkylate phosphorus by reaction with diethylzinc in ether solution to form tertiary phosphine (**27**), the first to contain a heterocyclic ring (Eq.9).

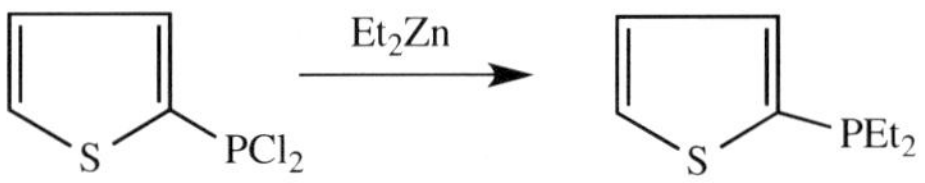

16 **27**

(9)

Other synthetic work included:

(1) The first synthesis of n-octylphosphine, using the Hofmann method of reacting phosphonium iodide with n-octyl iodide in the presence of zinc oxide in a sealed tube [32,33].

(2) The synthesis of dimethyl(p-tolyl)phosphine from dimethylzinc with p-tolylPCl_2 [34]. The phosphine was not oxidized on exposure to air, but could be oxidized with HgO to give the tertiary phosphine oxide.

The use of ether-soluble alkyl zincs was a definite advance in phosphine synthesis, but later was to be replaced by the use of the magnesium-based Grignard reagents.

(3) The first authentic synthesis of dibenzylphosphine. Letts and Blake, in a paper of 100 pages in length [12] reported an extensive study on the preparation and properties of mono-, di- and tribenzylphosphine. Hofmann had earlier performed the sealed-tube reaction of benzyl chloride with phosphonium iodide in the presence of ZnO to obtain three compounds to which he assigned these structures [35]. This reaction was repeated by Letts and Blake, and from the complex mixture of products they were able to isolate the three products of Hofmann. It was confirmed that Hofmann had indeed made benzyl and tribenzyl phosphines. It was found, however, that the compound isolated by Hofmann and thought to be dibenzylphosphine was probably tribenzylphosphine oxide. This paper is also of value in that it discusses in considerable detail the general properties of phosphines known up to that time.

Several other papers report on properties of the phosphine family of compounds. Two are of some interest. Holle [36] reported that the reaction of benzaldehyde with dimethylphenylphosphine in the presence of $AlCl_3$ gave a new type of product with structure **30** (Eq. 10). Diethylphenylphosphine also gave this type of product. Holle expressed the initial addition product as **28** with an (unlikely) 3-membered ring with 5-coordinate phosphorus (also unlikely), but a better representation would be the dipolar intermediate **29**. Holle pointed out the difference between an aromatic phosphine and an aromatic amine in this reaction; the latter would undergo attack on the benzene ring.

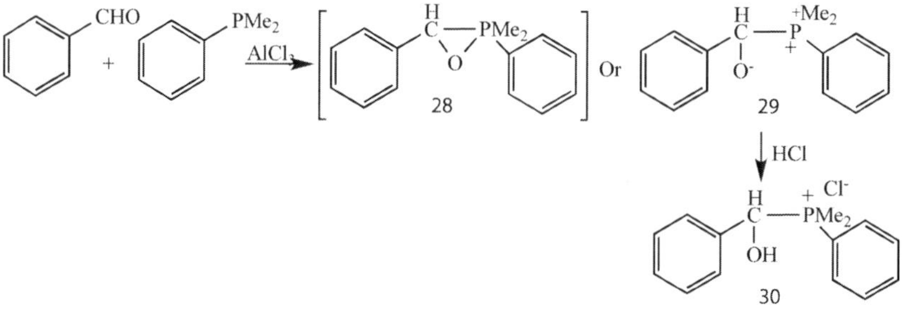

(10)

An unusual product was reported by Messinger and Engels [37] from the reaction of pyruvic acid with phosphine. The crystalline product on analysis was found to have the molecular formula $C_9H_9O_6P$ and thus was formed from three molecules of pyruvic acid and one molecule of phosphine with loss of three molecules of water. The authors wrote possible structures **31** and **32** for this compound; the correct structure (**33**) was discovered by S. A. Buckler in 1960 [38] using the spectroscopic techniques (infrared and NMR) not available in the time of Messinger and Engels. The product is therefore a heterocyclic tertiary phosphine, the first of a great many prepared in later years.

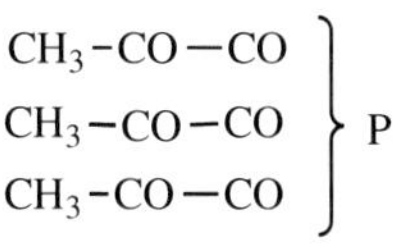

31

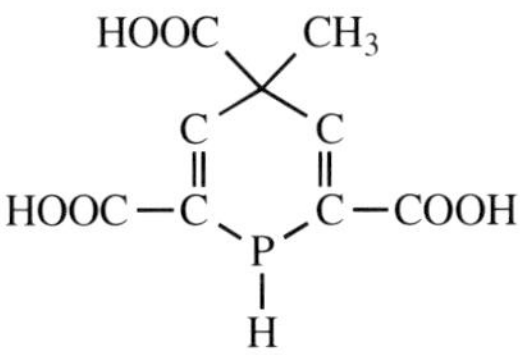

32

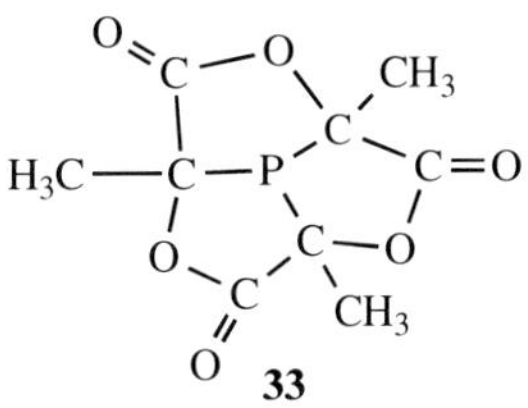

33

The field of coordination complexes with phosphines, now of enormous size and importance, had its beginning in the laboratories of the 19th century chemists. The first coordination complex of a phosphine with a metallic compound was prepared by Cahours and Hofmann in studies on triethylphosphine [39]. They isolated a crystalline product with $PtCl_2$ to which they assigned the structure $(Et_3P)_2$,HCl,$PtCl_2$ based on elemental analysis. The complex was later made by Cahours and Gal [40, 41] who noted that two isomeric forms were obtained, one white and insoluble in ether, the other yellow and ether soluble. Alfred Werner [42] in his pioneering work on geometrical isomerism among coordination compounds recognized that the complex $(Et_3P)_2PtCl_2$ existed as cis, trans isomers, the yellow form being cis and the white form trans, and assigned the correct structures (**34** and **35**, respectively), to these isomers.

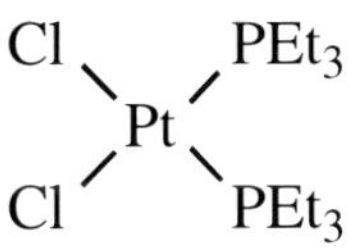

34, cis, yellow

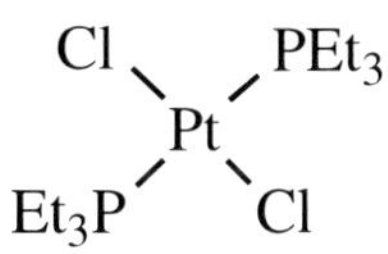

35, trans, white

4.4. Phosphine Derivatives: Salts and Oxides

As the chemistry of tertiary phosphines was being explored, it became the practice to characterize them as phosphonium salts or as phosphine oxides. These are usually crystalline derivatives, although frequently hygroscopic. They are easily formed in high yield, and without the terrible odor associated with many free phosphines. Most of the triaryl, trialkyl, and mixed alkylarylphosphines discussed in the preceding section on phosphines were converted to these derivatives, which need not be described here as they do not represent any additional novelty. However, the extension of phosphonium salt synthesis from two novel types of phosphines can be noted. In the heterocyclic area, Sachs [20] reported the methiodide (**36**) of (diethylphosphino)thiophene, and Kelbe [25] reported the ethiodide (**37**) of the first phosphine with polycyclic structure. Another novel salt (**38**) was made by Gleichmann [43] from the reaction of ethylene dibromide with dimethylphenylphosphine. He also prepared the diphosphonium salt (**39**).

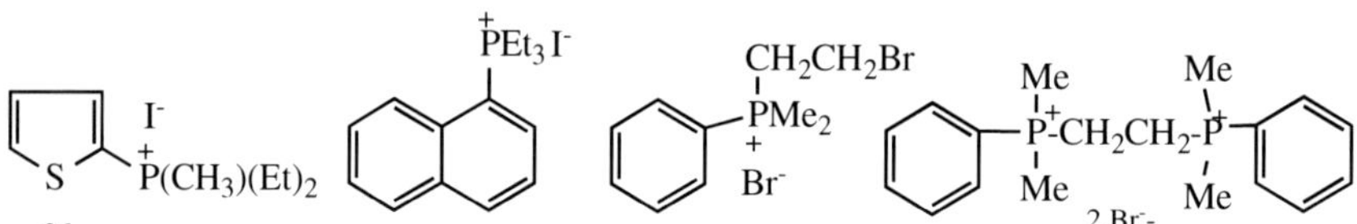

36 **37** **38** **39**

Tetrabenzylphosphonium compounds (**40**) were synthesized by Ledermann [44] by using the Hofmann method of heating phosphonium iodide with excess benzyl alcohol at 100°.

40

(11)

In Chapters 2 and 3, the work of Hofmann and Michaelis, respectively, on carboxyalkyl phosphonium salts was presented. After treatment with a base, these can be described as internal salts and called betaines (e. g., **41** from triethylphosphine and chloroacetic acid).

$$(C_2H_5)_3\overset{+}{P}\text{-}CH_2COO^-$$

41

As mentioned in Chapter 3, Meyer [45] made a new homolog of the betaine family by reacting trimethylphosphine with chloroacetic acid, followed by neutralization with base, and pointed out that the betaine was the phosphorus equivalent of the naturally occurring nitrogen counterpart. Conen [46] extended an approach of Michaelis to give new betaines in the aryl series (Scheme 1). He prepared 2,4-dimethylphenylphosphonous dichloride (**42**) as had been done by Weller [21], then methylating it with dimethylzinc to give the tertiary phosphine **43**. The methiodide (**44**) was prepared and converted to the corresponding chloride. This compound was subjected to oxidation by potassium permanganate which formed the monoacid of assumed structure **45**; in a separate reaction the second methyl group was oxidized to give the diacid **46**. With base, these compounds gave internal salts **47** and **48**, respectively, considered to be betaines, which Conen drew with a 5-membered ring from the carboxylate ion interacting with the phosphonium center. We now know that the true structures would simply be the dipolar forms **47** and **48**, as is true by definition for all betaines.

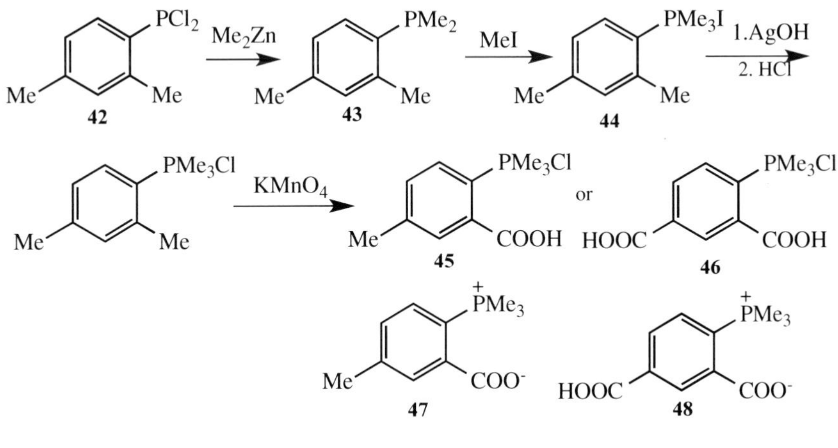

Scheme 1.

An important new synthetic method giving tetrahydroxyalkyl phosphonium salts was discovered by Messinger and Engels [37]. They found that phosphine introduced as the gas would not add to the carbonyl group of aldehydes as ammonia does, but if a stream of HCl gas was passed through the mixture, addition of P-H bonds to the carbonyl group did occur. Four molecules of the aldehyde became attached to phosphorus, giving salts of structure **49**. Various aldehydes were used in this reaction (Eq. 12), which became of modern interest when it was discovered that tetrahydroxymethylphosphonium chloride (known commercially as THPC), formed by the use of formaldehyde as reactant, imparted flame resistance to cotton fabrics.

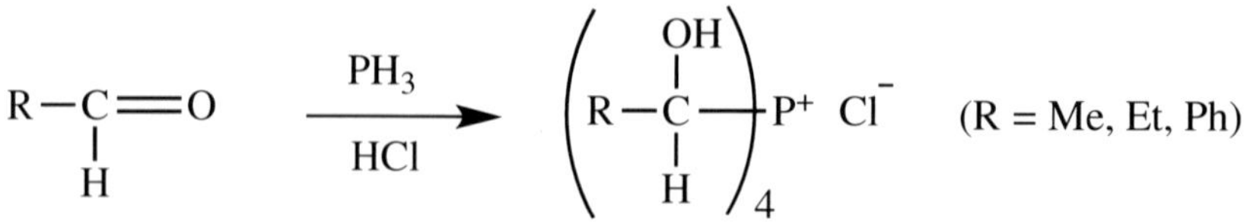

49

(12)

Prior to the Messenger-Engels paper, A. de Girard [47] had also described a reaction giving tetra(hydroxyalkyl)phosphonium salts. He reacted aldehydes with phosphonium iodide to obtain these compounds. This procedure has not attracted significant attention as has the reaction with phosphine and HCl.

Another new reaction that produced phosphonium salts was reported by P. Fireman, of the Catholic University of America, Washington, DC [48], one of the very few American investigators to contribute to nineteenth century phosphorus chemistry (another is A. H. Meyer from St. Louis, MO, writing on phosphonic betaines as noted in reference [45]). Fireman's paper is more of interest for indicating the international spreading of phosphorus chemistry, than for its chemical content. Fireman reacted phosphonium iodide with methyl and ethyl ether in a sealed tube in the range of 120-160°, obtaining a mixture of the tertiary phosphine and the tetraalkyl phosphonium iodide. The yields of these products were not reported, and the process does not seem to have attracted attention.

That the reaction of elemental phosphorus with ethyl iodide in a sealed tube could give compounds with C-P bonds was first observed by Carius [49] and by Crafts and Silva [50] who isolated some triethylphosphine oxide from the reaction mixture. This reaction was later clarified and developed by Masson and Kirkland [50], using different conditions, into a procedure for synthesizing tetraethylphosphonium iodide. The reaction was said to follow the stoichiometry of Eq. 13, employing a different reactant ratio than used by the previous workers. The iodide ion was converted to polyiodides in the process, and these were reduced to iodide ion in a subsequent reaction with ethanol.

$$2P + 7\,C_2H_5I \longrightarrow P(C_2H_5)_4I_3 + P(C_2H_5)_3I_4 \xrightarrow{C_2H_5OH} P(C_2H_5)_4I + PO(C_2H_5)_3 \quad (13)$$

Reactions using phosphorus as a reactant have become of major importance in recent years (being adopted by those performing "green chemistry" since phosphorus halides are not used as starting materials), and the Masson-Kirkland paper especially should be credited as a significant contribution to nineteenth century phosphorus chemistry.

In the Masson-Kirkland work [51] with tetraethylphosphonium iodide, it was observed that bromine or chlorine could add to the iodide ion and create a new "perhaloid" ion, as in IBr_2^- and ICl_2^-. The phosphonium chloride also added chlorine to form Cl_3^-. The salts had some stability and were of high color. Haloid ions were also formed with the sulfate ion, but their structure is uncertain.

The work of Norman Collie [28] on the decomposition of phosphonium salts bearing an ethyl group on phosphorus was presented in the section on phosphine synthesis. In later work [52] he studied the action of heat on various salts of tetramethylphosphonium ion. A peculiar reaction occurred with the chloride; at temperatures over 300° ethylene was formed and convincingly confirmed. Collie suggested that the decomposition released methylene (CH_2), perhaps one of the earliest mentions of this carbene in organic chemistry, as a reactive intermediate, which dimerized to ethylene. The other product was Me_2HPCl_2. Several other salts were similarly decomposed; those bearing oxygen, such as the benzoate, gave rise to trimethylphosphine oxide.

Very few significant advances were made in the synthesis and properties of phosphine oxides after their discovery by Hofmann and by Michaelis. Collie [53] made some useful remarks about the family in 1889, as being "a somewhat remarkable class of compounds. They are amongst the most stable substances to be met with in the field of organic chemistry". Collie goes on to say that they do not react with various potent reagents, yet they unite with acids, such as hydrochloric or hydrobromic acids to form well-characterized compounds, and with metallic halides to form "double salts". In particular, tribenzylphosphine oxide was prepared and became a focal point for study of the salts in the 1880's. It was first prepared in 1880 by Fleissner [54] using the Hoffman method of reacting benzyl chloride (presumably generated *in situ* from benzal chloride as a reactant by reduction) with phosphonium iodide and zinc oxide. He characterized it with the formation of several double salts with chlorides of platinum, palladium, iron, mercury, and cobalt. It was reported again by Collie in 1881 [55], by Siegfried Litthauer [56], and in an extensive study by Letts and Blake [12]. In the latter paper, many double salts were studied, as were the adducts with HCl and HBr. These authors did not express the hydrohalides with definite structures, being concerned about their analyses which did not reveal a definite 1:1 ratio of oxide to acid as would suggest a salt. The modern description depends upon knowledge that the P=O group does have basic properties, somewhere between that of the carbonyl group and the amine oxide group, and it is a strong receptor in hydrogen bonding. The products with acids may be the result of hydrogen bonding to HX molecules, or may reach the extreme of complete hydrogen acceptance thus forming salts [57].

Both Dörken [31] and Collie [53] reported that the aryl-substituted phosphine oxides could undergo typical aromatic substitution reactions. Dörken prepared the trinitro derivative (**50**) from the nitration of diphenylbenzylphosphine oxide. He did not draw a structure but it may be presumed that the nitro group entered the meta-position of the benzene substituent while it entered the para position of the benzyl ring. Collie prepared the trinitro derivative (**51**) from tribenzylphosphine oxide [54] as well as the trisulfonic acid (**52**) from sulfonation of this phosphine oxide. He showed that the former had the para-nitro structure by oxidation with potassium permanganate, which cleaved the C-P bond and importantly gave the known p-nitrobenzoic acid. Another new phosphine oxide, dimethyl(p-tolyl)phosphine oxide (**53**), was prepared by L. Czimatis [34] from the action of dimethylzinc on p-tolylphosphonous dichloride, followed by HgO oxidation of the resulting tertiary phosphine.

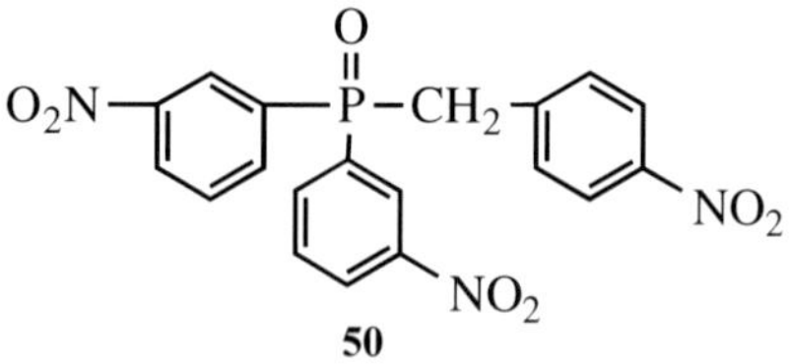

50

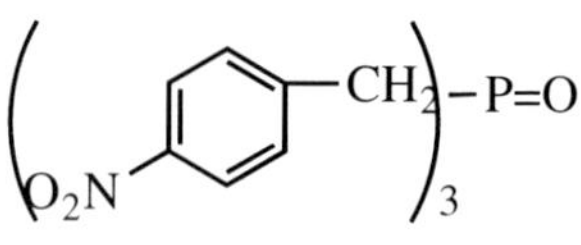

51

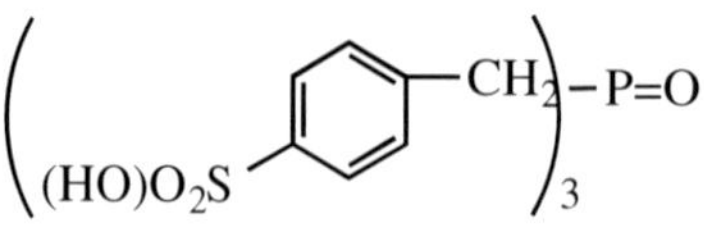

52

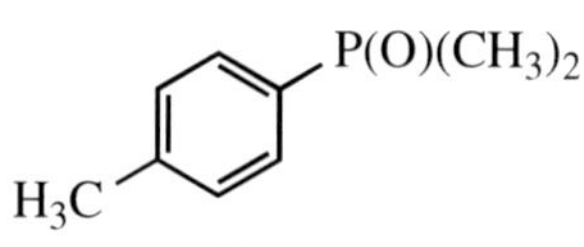

53

Tertiary phosphine sulfides received scant attention during this period. Dörken [31] did report that di(p-tolyl)phenylphosphine added sulfur in CS_2 solution to form the sulfide (**54**).

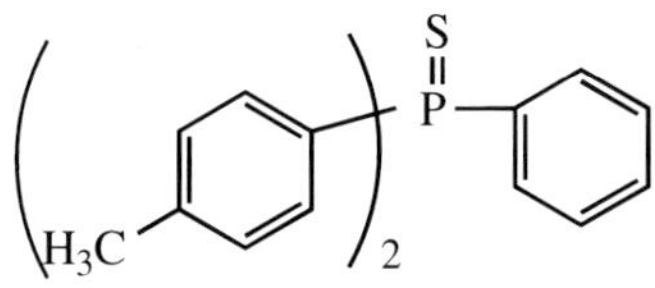

54

4.5. Phosphorus Halides

The research in the field of phosphonous dihalides (RPX_2) was quite limited during the nineteenth century, but of considerable importance was the work of F. Guichard, who synthesized the first alkyl phosphonous dichlorides [58] by the reaction of alkyl mercury compounds with PCl_3. His work has been discussed in Chapter 3. No other papers were concerned with the alkyl series of compounds. Aryl phosphonous dihalides attracted more attention. No new synthetic methods were devised, but several new compounds were made during this period using the procedures of Michaelis. Some have already been mentioned since they were prepared as precursors to other compounds, especially phosphonic acids, and will only be summarized here. Thus Sachs prepared the first heterocyclic compound, thienylphosphonous dichloride [20] by the PCl_3-$AlCl_3$ method, and naphthylphosphonous dichloride was prepared by W. Kelbe [23, 25] from dinaphthyl mercury on reaction with PCl_3. Weller [21, 22] prepared phosphonous dihalides by the PCl_3-$AlCl_3$ method from m- and p-xylenes; this method also was used by P. Melchiker [26] for the synthesis of 3-chloro-4-methylphenylphosphonous dichloride.

Dialkylphosphinous chlorides (R_2PCl) remained unknown during the nineteenth century, but diphenylphosphinous chloride was the subject of several reports. It was first mentioned in 1877 by Michaelis, in Carlsruhr *(sic)* at that time, as resulting from the reaction of phenylphosphonous dichloride with diphenylmercury [59]. In the next paper in the journal [60], A. Broglie also in Carlsruhr reported that this compound was formed by the pyrolysis of phenylphosphonous dichloride at 280° (Eq. 14). It is suggested that Broglie was associated with the Michaelis laboratory in Carlsruhr, but neither author mentioned the other. Michaelis and Gleichmann [61] later described the same pyrolysis reaction in detail, and it was employed by Dörken [31] in 1888 to prepare the material for studies on its properties.

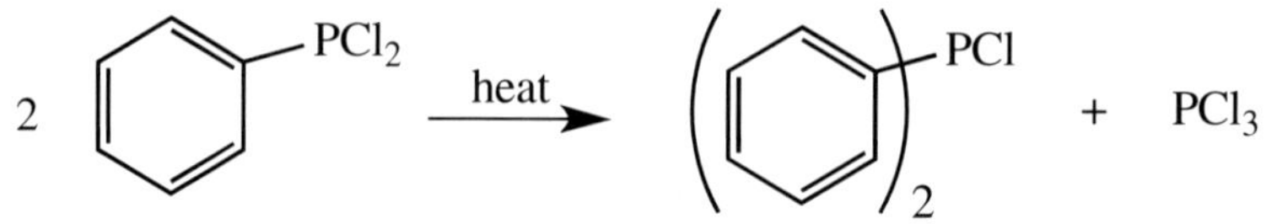

(14)

A significant report from Dörken's study, which amplified with more experimental data an earlier but brief observation by Michaelis and La Coste (given in Chapter 3) [62], was that diphenylphosphinous chloride could be alkylated by benzyl chloride to form a salt, in the same manner that a tertiary phosphine could be alkylated. The product was then hydrolyzed to give a phosphine oxide (Eq. 15). The alkylation of a phosphorus halide, which would be of low nucleophilicity, is a rare event and of very limited scope. Here the presence of two carbon substituents adds to the reactivity of phosphorus, and benzyl halides are strong alkylating agents.

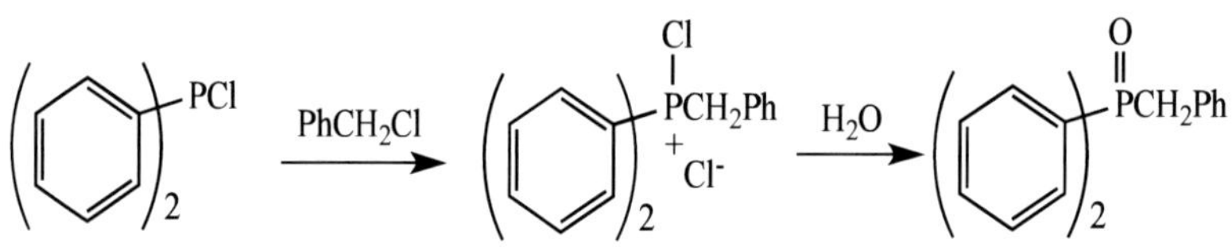

(15)

Dörken also repeated the synthesis of diphenylphosphine, first performed by Michaelis and Gleichmann [61], by heating a mixture of diphenylphosphinous chloride and sodium carbonate solution, whereupon a disproportionation reaction occurred with the formation of diphenylphosphine and sodium diphenylphosphinate. Dörken expressed this reaction as in Eq. (16).

$$4Ph_2PCl + Na_2CO_3 + H_2O = 2Ph_2PH + 2Ph_2PO_2Na + CO_2 \quad (16)$$

However, the process must occur in two steps, the first being hydrolysis to diphenylphosphine oxide (the preferred form in the tautomeric equilibrium with Ph_2P-OH, as discussed in Chapter 3) which is the substance undergoing the disproportionation. (Eq. 17).

$$Ph_2PCl \xrightarrow{H_2O} \left[Ph_2P\text{-}OH \rightleftharpoons Ph_2\overset{\displaystyle H}{\overset{|}{P}}{=}O \right] \xrightarrow{NaOH} Ph_2PH + Ph_2PO_2Na \tag{17}$$

Another result of Dörken was the replacement of the chlorine of Ph_2PCl with hydrogen, constituting a second synthesis of diphenylphosphine. As noted in the section on phosphines, this was accomplished by heating the chloride with metallic zinc at 230°, forming Ph_2PZnCl according to a chlorine analysis. On hydrolysis the phosphine was formed. Today we would perform such reductions much easier with metallic hydrides, $LiAlH_4$ in particular, but Dörken has the distinction of being the first to reduce a phosphorus halide.

Phosphonyl chlorides, $RPOCl_2$, were frequently prepared when phosphonous dichlorides became available. Examples will be found in the papers already cited by Sachs, Crafts and Silva, Fleissner, Fossek, Guichard, Kelbe, Melchiker, and Weller. These were prepared by oxidation or by addition of chlorine followed by reaction with sulfur dioxide, procedures developed by Michaelis (Chapter 3). Köhler reported that sulfur monochloride (known to be S_2Cl_2) acted on $PhPCl_2$ to form the thiophosphonyl chloride $PhP(S)Cl_2$ but the process has little utility [63].

Conclusion

Wide interest in organophosphorus chemistry developed in the final quarter of the nineteenth century, especially in Germany. The list of references given in Section 4.7 contains the names of about 45 chemists who made contributions to the field, large or small. Progress was made in the development of new synthetic methods for the functional groups discovered by Hofmann and Michaelis, and in the understanding of their properties. Use of the Structural Theory of Organic Chemistry was well in place and of great benefit. We are thankful to these early pioneers who dealt with difficult chemistry, and with some dangerous compounds having disgusting smells. With this background, the way was clear for the tremendous boom in phosphorus chemistry research and commercial applications that took place in the twentieth century, particularly in the second half.

REFERENCES

[1] Pebal, E. *Ann.* 1861, *120*, 194.
[2] Fossek, W. *Monatshefte* 1884, *5,* 121.
[3] Fossek, W. *Monatshefte* 1884, *5,* 627.
[4] Fossek, W. *Monatshefte* 1886, *7,* 20.
[5] Hofmann, A. *Ber.* 1872, *5,* 104.
[6] Ville, J. *Compt. rend.* 1888, *107,* 659.
[7] Ville, J. *Compt. rend.* 1890, *110*, 348.
[8] Ville, J. *Compt. rend.* 1889, *109,* 71.
[9] Ville, J. *Annales chim. phys.* 1891, *23,* 289.
[10] Litthauer, S. *Ber.* 1889, *22,* 2144.
[11] Kosolapoff, G. M. *Organophosphorus Compounds;* Wiley: New York, 1950, p 109.
[12] Letts, E. A.; Blake, R. N. *Trans. Roy. Soc. Edinburgh* 1889, *35,* 527.
[13] Behal, A. *Bull. Soc. Chim.* 1888, 632.
[14] Claus, A. *Ber.* 1881, *14,* 2365.
[15] Bishop, A. W.; Claisen, L.; Sinclair, W. *Annalen* 1894, *281*, 314.
[16] Michaelis, A.; Flemming, A. *Ber.* 1901, *34,* 1291.
[17] Marsh, J. E.; Gardner, J. A. *J. Chem. Soc. Trans.* 1894, *65*, 35.
[18] Gardner, J.A.; Cockburn, G. B. *J. Chem. Soc. Trans.* 1897, *73*, 704.
[19] Gardner, J. A.; Cockburn, G. B. *J. Chem. Soc. Trans.* 1897, *71*, 1156.
[20] Sachs, H. *Ber.* 1892, *25*, 1514.
[21] Weller, J. *Ber.* 1887, *20*, 1718.
[22] Weller, J. *Ber.* 1888, *21*, 1492.
[23] Kelbe, W. *Ber.* 1876, *9*, 1051.
[24] Lindner, J.; Strecker, M. *Monatsh.* 1929, *53*, 274.
[25] Kelbe, W. *Ber.* 1876, *11*, 1499.
[26] Melchiker, F. *Ber.* 1898, *31*, 2915.

[27] Michaelis, A.; Panek, C. L. *Liebigs Ann.* 1882, *212,* 203.
[28] Collie, N. *J. Chem. Soc. Trans.* 1888, *53*, 714.
[29] Letts, E. A.; Collie, N. *Phil. Mag*. 1886, *22*, 183.
[30] Drechsel, E.; Finkelstein, E. *Ber.* 1871, *4*, 342.
[31] Dörken, C. *Ber.* 1888, *21*, 1505.
[32] Möslinger, W. M. *Ber.* 1876, *9*, 1005.
[33] Möslinger, W. M. *Ann.* 1877, *185*, 49.
[34] Czimatis, I. *Ber.* 1882, *15*, 2014.
[35] Hofmann, A. W. *Ber.* 1872, *5,* 100.
[36] Holle, O. *Ber.* 1892. *25*, 1518.
[37] Messinger, J.; Engels, C. *Ber.* 1888, *21,* 326, 2919.
[38] Buckler, S. A. *J. Am. Chem. Soc.* 1960, *82,* 4215.
[39] Cahours, A.; Hofmann, A. W. *Ann.* 1857, *104*, 1.
[40] Cahours, A.; Gal, H. *Compt. rend.* 1870, *71,* 208, 897.
[41] Cahours, A.; Gal, H. *Compt. rend.* 1870, *70,* 1380.
[42] Werner, A. *Z. anorg. Chem.* 1893, *3*, 318.
[43] Gleichmann, L. *Ber.* 1882, *15,* 198.
[44] Lederman, B., *Ber.* 1882, *21*, 405.
[45] Meyer, A. H. *Ber.* 1871, *4*, 734.
[46] Conen, F. *Ber.* 1898, *31*, 2919.
[47] de Girard, A. *Annales chim. phys.* 1884, *2*, 1.
[48] Fireman, P. *Ber.* 1897, *30*, 1088.
[49] Carius, L. *Ann.* 1866, *137*, 117.
[50] Crafts, J. M.; Silva, R. *J. Chem. Soc.* 1871, *24*, 629.
[51] Masson, O.; Kirkland, J. B. *J. Chem. Soc. Trans.* 1889, *55*, 126.
[52] Collie, N. *J. Chem. Soc. Trans.* 1888, *53*, 636.
[53] Collie, N. *J. Chem. Soc. Trans.* 1889, *55*, 223.
[54] Fleissner, F. *Ber.* 1880, *13*, 1665.
[55] Collie, N.; Letts, E. A. *Trans. Roy. Soc. Edinburgh* 1881, *11*, 46; *J. Chem. Soc. Trans.* 1882, *42*, 724.
[56] Litthauer, S. *Ber.* 1889, *22*, 2144.
[57] Edmundson, R. H. *The Chemistry of Organophosphorus Compounds, Vol. 2*; John Wiley: Chichester, UK, 1992, p 289.
[58] Guichard, F. *Ber.* 1899, *32*, 1572.
[59] Michaelis, A. *Ber.* 1877, *10,* 627.
[60] Broglie, A. *Ber.* 1877, *10*, 628.
[61] Michaelis, A.; Gleichmann, L. *Ber.* 1882, *15*, 801.
[62] Michaelis, A.; La Coste, W. *Ber.* 1885, *18,* 2109.
[63] Köhler, H. *Ber.* 1880, *13*, 463.

APPENDIX: FULL NAMES OF JOURNALS CITED

Annalen der Chemie und Pharmacie
Annalen der Chemie und Pharmacie. Supplementband
Annales de chimie et de physique
Berichte der deutschen chemischen Gesellschaft
Bulletin de la Société chimique de Paris
Comptes rendus hebdomadaires des séances de l'Académie des sciences
Gazzetta chimica Italiana
Journal of the Chemical Society
Journal of the Chemical Society, Transactions
Journal für praktische Chemie
Monatshefte für Chemie und verwandte Teile anderer Wissenschaften
The London, Edinburgh and Dublin philosophical magazine and journal of science.
Transactions of the Royal Society of Edinburgh
Zeitschrift für anorganische Chemie

INDEX

A

B

C

D

E

F

G

H

I

K

L

M

N

O

P

Q

R

S

T

V

W